Transforming Business with Program Management

Integrating Strategy, People, Process, Technology, Structure, and Measurement

PRAISE for the BOOK

"Satish offers a practical guidebook that balances both the art and science of program management—regardless of what industry you are in. It has the credibility that only comes from someone who has been there. Well done!"

F. WARREN McFARLAN
Baker Foundation Professor, Harvard Business School

"This book is a must read for those involved with large-scale transformation programs...and today, that's most everyone, including senior executives. It seems like every company faces a major transformation every few years, driven by the ever increasing pace of change and new competitive pressures caused by mega-trends such as digitization and globalization. Running these transformation programs well is a competitive advantage in itself, and Satish's book is a comprehensive yet pragmatic view of how to make these transformation programs successful."

ROMIL BAHL
Executive Vice President and General Manager, Global Industries, CSC

"Satish does an excellent job of telling readers how program management can help transform companies. As the primary research analyst in the design of the 60 points of transformational leadership, I instantly recognized the credibility of how Satish details the importance of people and talent in his six principles of transformational program management. It was a pleasure to read a book written by a true expert and peer who supports many of the theories we have found to be accurate in the years of research we have done at CTPartners and Transformational Labs on the relationship between transformational talent and business transformation. Satish simplifies what many of us embrace as a complex task of assembling the right talent required for effective transformational program management. Certainly a must read for anyone wanting to grasp an in depth knowledge of this topic or tasked with the implementation of program management that has the ability to transform thinking and results."

JEFF CHRISTIAN
CEO and Founder, Transformational Lab
Former CEO and Founder, CTPartners (formerly Christian & Timbers)

"Satish has introduced a pragmatic approach which is long overdue in the transformation business. He obviously has been involved in a number of successes and failures which only makes his insights richer and more meaningful. If you have not driven transformation at this level then this book is your primer—if you have than this will serve to help up your game next time around."

JIM CARUSO
Executive Vice President, World Wide Services, EIS Group

"This book serves as an excellent resource for program, project, portfolio, and change managers, providing detailed coverage of 15 techniques that need to be applied to drive large scale programs and projects to success. Satish follows a clear and distinct path—each chapter unfolds with precision and delivers tangible tools and steps to implement. If you appreciate clarity, Satish has written this book for you. Reference the road map in his book for managing strategic initiatives, cross-functional programs and complex projects, and learn from a senior practitioner who has been there, done that...and written about it!"

BILL YATES
Executive Vice President, Velociteach

"This is a genuinely new and exciting look at program management. It takes the professional discipline of program management and demonstrates its practical value in the context of a complex transformation environment. A 'must read' for the C-suite!"

GLOBAL PORTFOLIO AND PROGRAM MANAGEMENT LEADER
Big Four Consulting Firm

"Managing a large complex portfolio of critically important programs and projects has never been easy. Yet the skills required have never been more important as businesses everywhere work to constantly reinvent themselves in response to rapidly changing global markets. Without a comprehensive framework to guide the program manager, the accumulated bodies of knowledge available will often overwhelm rather than help ensure success. In "Transforming Business with Program Management", Satish Subramanian brings just the right blend of academic rigor, logic, and a practitioner's pragmatic approach to the description of such an essential framework. Covering all the necessary aspects—strategic, procedural, technical, and human factors—Satish sets out to describe and organize all the things a program manager must ensure happen to deliver transformational results, providing a valuable handbook for anyone setting out to manage transformational change."

JOHN PARKINSON
Chairman and Managing Director, Parkwood Advisors

"This book provides a clear roadmap for business organizations to successfully implement the kind of sophisticated and integrated transformation so often needed in today's fast-changing global economy. It is well researched, providing a wealth of conceptual knowledge and case studies and it provides a concrete framework for the hardest part of strategic market adaptation, getting the execution right. I urge business leaders to read it now, before the competition does."

DR. CHARLES MAXEY, Ph.D.
Founding Dean, California Lutheran University of School of Management
Co-Director, Center for Leadership & Values and Fellow, Center for Economic Forecasting & Research
California Lutheran University

"Satish Subramanian's book is a must-read for anyone planning a new, or struggling with a current, complex business-critical initiative. The book is highly structured, easy to read, and provides valuable models, processes, techniques, and tools for managing large-scale programs. Satish's insights and methods are then described in real-world situations through numerous case studies. The book is not only a practical roadmap for practitioners to successfully deliver the expected business outcomes, but it is also a comprehensive textbook for learning about program, project, and organization change management. Everyone in the business and educational sectors would benefit from reading this book and taking Satish's insights to heart."

SAM DOYING
Sales Executive, Big Four Consulting and High Tech Industry

"Effecting transformational change in any organization comes down to the successful coordination of many discrete yet related strategic initiatives. This book provides a powerful template for success in business transformation and program management."

TIM Z. DILLEY
Cloud Computing Consulting Executive

"Satish provides observations, evidence, and instructions on program management objectives across business transformation efforts. Using the disciplined approach described in this book will increase the odds of success for strategic change management. Each sentence is packed with content that deserves the reader's thoughtful analysis in customizing the approach to achieve the program's objectives. Leaders are the catalyst for change and, by using the techniques in this book to engage company employees the organization becomes the agents of change. I've known Satish for almost 20 years and have seen him use the principles in this book to help companies in a variety of complex strategic initiatives; from merger acquisition integration, to systems implementation, to operational change, and regulatory compliance. The book's recipes are not just a how-to checklist; rather the book provides a framework to replicate the success Satish has proven in his life's work. I love the way Satish provides information to guide PMO's and manage change and recommend this book to anyone involved in a transformation program."

PARTNER and EXECUTIVE CONSULTANT
Big Four Professional Services Firm

Best Practices and Advances in Program Management Series

Series Editor
Ginger Levin

Transforming Business with Program Management

Integrating Strategy, People, Process, Technology, Structure, and Measurement

Satish P. Subramanian

Foreword by **Thomas H. Davenport**

CRC Press
Taylor & Francis Group
Boca Raton London New York

CRC Press is an imprint of the
Taylor & Francis Group, an **informa** business

CRC Press
Taylor & Francis Group
6000 Broken Sound Parkway NW, Suite 300
Boca Raton, FL 33487-2742

© 2015 by Taylor & Francis Group, LLC
CRC Press is an imprint of Taylor & Francis Group, an Informa business

No claim to original U.S. Government works

Printed on acid-free paper
Version Date: 20150128

International Standard Book Number-13: 978-1-4665-9099-1 (Hardback)

Visit the Taylor & Francis Web site at
http://www.taylorandfrancis.com

and the CRC Press Web site at
http://www.crcpress.com

Dedicated to my parents (Mangalam and P.H. Parameswaran),

my wife (Jaya), and our sons (Shasta and Tanay) who all have

been my sources of core values, strength, and confidence.

Contents

List of Case Studies

Foreword

For much of my working life, I have researched, written about, advocated for, or consulted about—and in a few cases even helped implement—the types of business change programs that are the focus of this book. In fact, I believe that I first met Satish Subramanian at Ernst & Young, where we were both trying to advance the art of business process reengineering.

I am still a big believer in the importance of such programs. No matter what their focus, they share one objective—making things better and creating better performance. I think that seeking improvement is perhaps part of the human condition, and one of the most exciting aspects of business. If you do the same job every day, business can be boring. But if you are constantly working to make the business better, it's a fascinating enterprise.

But these transformational change programs have a mixed record of success. In many cases, a senior executive gets excited about an idea after reading about it or hearing it presented at a conference. He or she charges a team with implementing the idea, often with the help of consultants. Implementing the change typically gets much less ongoing attention than normal operations and strategy execution.

So I welcome the level of attention that organizational program management is beginning to get from authors like Satish Subramanian and books like this one. Systematic attention to all the key issues of these programs can make them much more likely to succeed. If we're going to take on transformational change programs, it's only logical to manage them in a way that greatly improves their impact and effectiveness.

I am confident that Satish has identified each of the key steps toward effective program management in his road map. I like that he starts with establishing clarity about the nature of the problem—I call it "framing the issue"—and proceeds all the way through the transition to a fully implemented new state. And he considers all the change tools you will need along the way—from strategy to technology, and all the human change issues as well.

The models in this book are somewhat complex, but then so is transformational change. One of the reasons why these programs fail is that a leader fails to attend to some critical aspect of the change—the explanation of the need for change, the education and training of front-line workers, the technology, or something else. This book can serve as a checklist for change leaders to help them ensure they are covering all the bases and taking all the necessary steps.

There is already plenty of advice in Satish's book, and I don't want to add much. But I will say one thing that may be of use. Given the amount of managerial effort involved in successfully managing transformational change programs, you probably want to limit the number of them. GE under Jack Welch, for example, was one of the best companies at managing change programs I have ever seen. In addition to a "Strategic Initiatives" organization and a very active involvement level by Welch in poking and prodding business units to make necessary changes, the company strictly limited the number of concurrent change programs. Six Sigma, services orientation, digitization—these were all considered great ideas, but they all required substantial labor and attention to address successfully.

So GE limited the number of programs to five or six, even though it was one of the world's largest corporations and had a high ability to transform itself when necessary. If GE could handle that many, most organizations should have substantially fewer programs underway at once—perhaps as few as one or two. My own experience and a few informal surveys, however, suggest that large organizations often have as many as twenty different change programs going at once. No matter how good your methods and tools, this is too many to achieve success.

So choose your shots carefully, and employ all the tools herein to make them winning programs. Whether you read this book and internalize its content, or—better yet—give it a quick scan and then use it as a reference source as you manage change programs, I am confident that you will find it very helpful. Paying attention to all the details that Satish Subramanian points out will make it easier to focus on the exciting, inspirational aspects of business change. For many of us, it's why we were attracted to business, and why we keep at it.

Thomas H. Davenport
President's Distinguished Professor of IT and Management, Babson College
Research Director, International Institute for Analytics
Digital Fellow, MIT Center for Digital Business

Preface

This book began with the objective that there is an enormous need to start the conversation on the integration and linkages among business strategy execution, business transformation, program management, and organization change management. The premise is that business performance can be substantially improved by recognizing, understanding, and capitalizing by integrating and aligning these disciplines. Organizations need to constantly innovate and improve products and services to maintain a strong competitive position in the market place. The vehicle used by organizations for such constant reinvention is a business transformation program. Such a program can be executed from start to finish through the successful application of the end-to-end program management life cycle. The focus of this book is on detailing how program management delivers the tangible measurable business outcomes and sustains the desired business change brought about by the successful completion of business transformation.

WHY I WROTE THIS BOOK?

1. To exhibit that program management capability bolsters the competitiveness of organizations as it successfully drives disruptive changes, which results in innovations and radical improvements.
2. To establish the linkage of program management to strategy execution and how it integrates the other needed disciplines to realize and sustain the outcomes expected from strategic initiatives.
3. To demonstrate proven and practical strategies, frameworks, and techniques that need to be adopted by organizations to improve the success rates of business transformation programs.

HOW IS THIS BOOK DIFFERENT?

The book lucidly elaborates how program management integrates and aligns the six dimensions or disciplines (strategy, people, process, technology, structure, and measurement) that are needed to successfully transform the business, execute strategic initiatives, and sustain the outcomes.

1. The book spells out the holistic program management approach adopted by enterprises to successfully execute strategies that improve their competitive positioning.
2. This practical resource elevates the significance and visibility of program management function within an organization by detailing how it enables realization of vision and objectives of business transformation.
3. The handbook defines and expands on the key techniques that need to be part of any program management toolkit.

WHAT CONTENT DOES THIS BOOK COVER?

The twelve chapter book provides the reader with a tested program management road map along with the supporting comprehensive frameworks to successfully execute transformation programs, formulated strategies, and strategic initiatives. It describes how program management integrates multiple disciplines in enabling organizations to attain the intended strategic objectives and deliver tangible business outcomes. This "how to guide" highlights what organizations need to do to enhance their strategic execution capability and exhibits ways to improve their program management maturity, which mitigates the risk of failure of business transformation. The book outlines the ten essential steps to successfully transform any business through effective execution of the program designed to transform the business.

This breakthrough work establishes the linkage between strategy formulation and strategy execution through the program management discipline. This insightful text equips executives and practitioners with the core skills necessary to effectively plan and implement business transformation strategies that drive sweeping business change and innovation. This book showcases processes, techniques, and tools that a program management team can customize and easily implement on any type of strategic initiative within the private or public sector environment to deliver and sustain the expected business outcomes and benefits.

HOW WILL THIS BOOK BENEFIT THE READER?

1. In addition to providing proven customizable frameworks, this text provides numerous templates that can be quickly leveraged to lead complex programs to success.
2. Illustrated with numerous figures, the book is a practical on the job resource that showcases business transformation and program management best practices, captures the lessons learned, and spells out the critical success factors for executing strategic initiatives and complex cross functional programs.
3. This "how to guide" provides expert advice on establishing program management center of excellence that delivers and sustains the intended outcomes of business transformation.
4. Through real world case studies within organizations spanning multiple industry sectors, this handbook provides executives, practitioners, and students with a guided tour of the program management operating model, program architecture, and program management life cycle.
5. The book integrates business and technology perspectives as that is one of the critical success factors for effective program management in today's fast evolving global market place which necessitates greater application of technology to meet the rapidly changing customer needs.

WHO IS THE TARGET AUDIENCE FOR THIS BOOK?

1. Business and technology personnel at companies in any industry sector who are involved in programs, projects, portfolios, or initiatives of transformational or nontransformational type.
2. Executives, general managers, and practitioners at companies of all sizes with the ownership and responsibility for portfolio management, program management, project management, and organization change management within a functional area or across functions.
3. Portfolio managers, program managers, project managers, or organization change specialists at public sector organizations as the principles covered in the book equally apply to them too.

4. Strategists focused on formulating strategies that are feasible from an execution perspective.
5. The ten real world case studies, sixty one illustrations, and inclusion of the latest research in the book, lends it to be a good text for academicians teaching various graduate level courses.

I hope you find this book interesting, derive much value in reading it, and benefit by applying what you learned from it.

Satish P. Subramanian
satishps@outlook.com
http://www.linkedin.com/in/satishsubramanian
San Francisco, California
January, 2015

Acknowledgments

As this book reflects an amalgamation of more than twenty-five years of my work, it is impossible to name and adequately express my appreciation to all the individuals who have shaped my thoughts, offered me advice, and guided me over the years. There are many acknowledgments due when a book represents so many years of experience. In addition, there are so many people to thank who have provided me with support and encouragement in writing this book. I am deeply grateful to Ginger Levin for her mentorship and guidance throughout my book journey.

I have been very fortunate to have had the opportunity to drive, manage, and work on business transformational initiatives launched by many companies. I am deeply indebted to my current employer (M Squared Consulting, a SolomonEdwards Company) and prior employers (Ernst & Young, Infosys, Point B, Cambridge Technology Partners, and Godrej & Boyce) who have helped me grow professionally. I am thankful, grateful, and appreciative of all my employers for providing me numerous opportunities to improve the business performance of companies.

I am deeply thankful to all my clients, work colleagues, teams, advisors, and professors who have made this book project a learning, rewarding, and meaningful experience for me and for the chance to share this knowledge. In addition, I sincerely appreciate the periodic opportunities from Global Knowledge, California Lutheran University, and Golden Gate University to engage with many experienced practitioners and graduate level students who have influenced my ideas.

It has been a pleasure to work with my publisher, CRC Press. I would like to give a special thanks to John Wyzalek for his direction, insights, and guidance. In addition, I would like to make a special note of the editing work of Prudy Taylor Board and the early set ups by Amy Blalock. I want to acknowledge the diligent efforts of the entire CRC team in getting this book published.

I wish to recognize all those people who have participated in a successful initiative to transform the business and those currently helping to make business transformation initiatives a reality in their own companies and organizations. The contributions of all individuals in progressing our collective understanding of business transformation, program management, and organization change management is appreciated as that helps us in increasing the success rate of complex programs and large projects with either a transformational or nontransformational agenda.

Finally, my greatest appreciation is reserved for my immediate family members. I want to thank my parents who toiled hard to get me educated and allowed me the opportunity to pursue my passion. I could not have accomplished this work without the tremendous support of my wife who took care of most of the family's needs as I worked through late nights, early mornings, and weekends. I am grateful to my children for their understanding of dad missing some of their events and curtailing play time with them for several months.

Thank you.

Satish P. Subramanian
San Francisco, California
January, 2015

About the Author

Satish P. Subramanian is a principal at M Squared Consulting, which is a SolomonEdwards Company. His 25 plus years of management and technology consulting leadership experience has resulted in tangible value creation for F500 and global companies in the health care, financial services, high tech, and manufacturing industry sectors. He has advised companies on their business transformation initiatives, guided them on operational optimization, driven strategic change to realize vision, executed strategic programs, and delivered sustainable results. He has built the business transformation and program management capabilities of organizations.

Satish has held executive-level positions at Ernst & Young, Infosys, Point B, and Cambridge Technology Partners. He has cultivated trusted advisory relationships with clients in delivering tangible business outcomes for companies of all sizes. He has orchestrated the realization and sustainment of growth and cost reduction objectives of organizations by applying a collaborative and solution driven approach. He developed his deep expertise through his hands-on consulting engagements, line role at a South Asian manufacturing conglomerate, and thought leadership in the form of authoring, speaking, and training. He is an adjunct instructor at Global Knowledge, a global company providing learning solutions and training services to companies.

Satish holds a BS in industrial engineering and MBA (Sales and Marketing) from the University of Mumbai and an MBA (Management Information Systems) from California Lutheran University. He is Program Management and Project Management certified by the Project Management Institute and his Change Management certification is from Prosci. His leadership contributions include speaking, training, and authoring. He invites you to share your thoughts on his book. He can be reached either at satishps@outlook.com or http://www.linkedin.com/in/satishsubramanian.

1

Executive Overview

The business world has changed radically in the past decade. Globalization, smart devices, and social media have all had a profound effect on how we approach work and get important programs done. For business transformation initiatives to succeed in this shifting environment, organizations today must give renewed emphasis to the tenets of program management, which provide the focus, structure, and discipline necessary to achieve desired business outcomes. In a fast evolving, flat world operating environment, enterprises need to constantly innovate and improve products and services to maintain a strong competitive position in the marketplace.

The vehicle used by enterprises for such constant reinvention is a business transformation program. Most U.S. based health care organizations have launched various transformation initiatives to enable them to get ready for the sea changes that are being brought in the health care industry by the health care reforms. In the financial services industry sector, the explosion of digitization has led financial institutions to initiate business transformation initiatives to handle the demand for new financial products and services. The disruptions from cloud computing technology is forcing highly successful and global technology companies to embark on business transformation to change the crux of their business models.

A holistic, structured, and rigorous program management practice is critical to making a business transformation happen and sustain. The program management discipline can integrate and align the six critical dimensions (strategy, people, process, technology, structure, and measurement) needed to transform a business through a transformation program. Business transformation programs can be executed from start (initiation) to finish (operational transition) through successful application of the end-to-end program management life cycle. The following topics are covered in the first chapter of the book:

- Book overview
- Road map to make a business transformation program successful
- Competitive advantage and program management framework
- Key techniques to facilitate successful business transformation
- Chapter synopses (Chapters 2–12)
- Transformation program management closes the business outcome gap
- Transformation program realizes business benefits
- Clarification on benefits realization terminology
- Business transformation and program management life cycle

BOOK OVERVIEW

This book enumerates how enterprises are applying program management to realize business benefits by successfully delivering the outcomes expected of complex, strategic, cross functional initiatives. The book will facilitate a better understanding and application of program management, a key enabler for improving the success rate of business transformation programs. The book synthesizes the principles of competitive strategy, organization change management, process improvement, automation through technology, organization structure, and measurement architecture through the program management umbrella. Strategists and execution specialists working on transforming a business will find this book invaluable. The book focuses on the following:

1. Details in a pragmatic way the ten mandatory steps (or road map) needed to lead complex, business transformation programs to success
2. Describes the holistic program management approach adopted by global enterprises to execute strategies that improve their competitive positioning
3. Elaborates how program management enables the fusion of strategy, people, process, technology, structure, and measurement on cross functional initiatives
4. Defines and expands on the key techniques that need to be part of any program management tool kit for a transformation program to realize the transformational change

5. Depicts how program management integrates a business and technology view, which is a critical success factor for implementing business transformational strategies

6. Serves as a how to guide for enterprises focused on building and enhancing their program management and transformation management competencies

7. Draws a distinction between program leadership and program management and the need for both to deliver the business benefits expected of a transformation program

8. Showcases program management best practices and lessons learned though real-world case studies spanning different industry sectors and functional domains

This book will appeal to business and technology professionals who are involved in some capacity on business transformation programs that are initiated to solve either complex business problems or improve business performance. Executives, general managers, and practitioners with responsibility for a program management office, portfolio management, program management, and project management will be able to leverage the key concepts and techniques covered in this program management book. The book will cater to the needs of the day-to-day tactical project managers and senior level decision making program managers who are on the business and technology side.

ROAD MAP TO MAKE A BUSINESS TRANSFORMATION PROGRAM SUCCESSFUL

The complexity of successfully delivering disruptive change that improves organizational performance through a transformation journey is high. In addition, the risk of not realizing the expected business benefits from a business transformation program is high. This book details the approach to be taken to drive a business transformation program to success. Program managers oversee the process of solving complex business challenges by driving end-to-end processes, managing project managers who oversee project work streams, and engaging with stakeholders to manage their expectations of program outcomes. True transformation program management—and, ultimately, program success—is best achieved by flawlessly executing the following road map:

- Success Starts Upfront: Describe the Problem Accurately
- Articulate the Program Vision and Objectives
- Secure Cross Functional Executive Sponsorship
- Develop and Implement a Governance Model
- Define Success, Outcomes, and Key Value Indicators
- Invest in Planning and Creating an Integrated Approach
- Drive Strong Partnership and Stakeholder Engagement
- Provide Leadership Across All Levels
- Monitor Aggressively and Have Contingencies
- Create and Implement an Operations Transition Plan

Each of the ten components of this road map for a business transformation effort to attain the desired future state objectives has been allocated a chapter in this book. Although the road map components are strategically sequenced, the planning and execution of the road map is an iterative exercise. A pictorial view of the proven road map to drive a business transformation program to success in the eyes of the stakeholders is presented in Figure 1.1. The program team utilizes the techniques in the program management tool kit to deliver in line with the road map. The program management team needs to have the capability (hard skills, soft

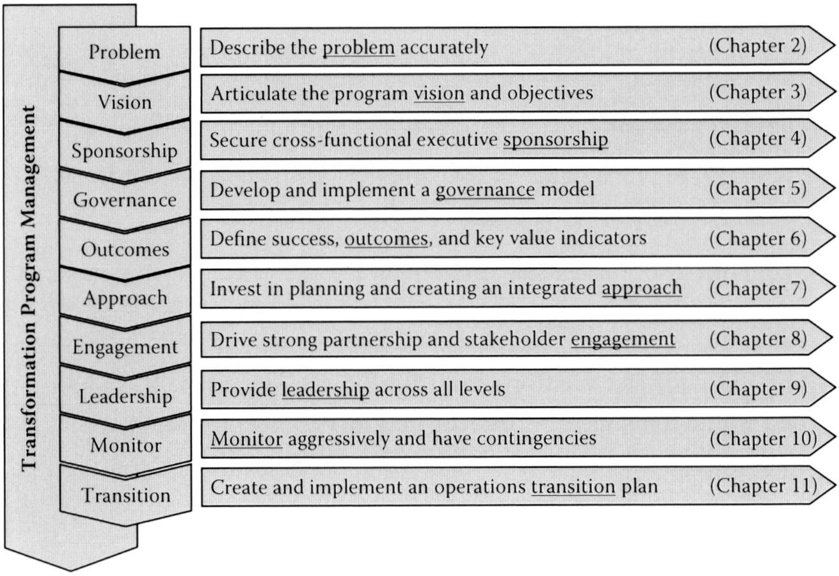

FIGURE 1.1
Road map for transformation program success.

skills, methods, processes, techniques, and tools) to execute against the road map.

COMPETITIVE ADVANTAGE AND PROGRAM MANAGEMENT FRAMEWORK

High performing enterprises, as well as those lagging the leaders, are always on the lookout for the levers that give them a business performance edge and are always striving to maintain that edge. A business strategy discipline has numerous frameworks and tools that are used by enterprises to formulate their competitive strategies. Enterprises recognize the equal importance of possessing the strategic execution capability needed to effectively implement a competitive strategy. This "must have" strategic execution capability is also the capability of program management.

Enterprises typically launch a business transformation initiative to achieve strategic business objectives, realize desired business outcomes, and improve their competitive positioning. According to research conducted among 587 C Suite and senior executives by the Economist Intelligence Unit (2013), organizations can improve their competitiveness by successfully executing initiatives to deliver strategic results. A transformation initiative could be structured either as a single large program or multiple programs, with the subprograms composed of numerous projects.

The multidimensional, integrative, and holistic framework of program management enables enterprises to initiate, plan, execute, and monitor a transformation initiative. The program management framework factors the six dimensions (strategy, people, process, technology, structure, and measurement) in delivering the desired business outcomes, realizing the business benefits, producing the tangible and intangible business results, and sustaining the benefits. A high-level explanation of the six dimensions follows:

1. *Strategy*: The transformation initiative undertaken by a program has to be aligned with the organization's strategy and the strategic objectives that need to be accomplished.
2. *People*: Program management entails engaging with stakeholders, managing human capital, and preparing the workforce to embrace the changes the business transformation is driving.

FIGURE 1.2
Program management framework.

3. *Process*: Typically, a business change developed by a complex program requires the redesign of current business processes and implementation of the redesigned new processes.

4. *Technology*: In light of the ever changing technologies and innovations they enable, most transformation programs involve supporting the changes to the technology landscape.

5. *Structure*: The current organizational model, role, and work location needs to be assessed and restructured to attain and sustain the future state goal of a strategic initiative.

6. *Measurement*: A measurement architecture has to be designed and implemented to monitor and evaluate the delivery and sustainment of business value expected from a transformation.

Based on the context, situation, and nature of the transformation program, it is possible that some of these six dimensions may require less (or more) work relative to the other dimensions during certain phases of program execution. One additional lever of competitive advantage for enterprises is building the program management competency (Figure 1.2).

KEY TECHNIQUES TO FACILITATE SUCCESSFUL BUSINESS TRANSFORMATION

This book provides an in depth coverage of fifteen techniques that can be applied during the course of a transformation program to successfully transform a business. For each of these fifteen techniques, the book presents an overview, an objective, and an approach for using the technique successfully. Knowledge of these techniques will enable program management practitioners to drive the complex business transformation programs that they are leading to success by delivering the business benefits and outcomes expected by the program stakeholders.

1. Environment scanning
2. Voice-of-customer
3. Strategic alignment
4. Business performance calibration
5. Program value justification
6. Governance modeling
7. Governance policy design
8. Performance improvement measurement
9. Business-outcome modeling
10. Program architecture
11. Organization change management
12. Transformation program planning
13. Stakeholder expectation management
14. Value-enhancement analysis
15. Walk through

Please note that only some of the key program management techniques have been expounded in this book and that there are many other techniques. In the next few pages, an executive level summary snapshot of each of the ten road map components is provided. The fifteen techniques in the previous list are spread across Chapters 2–11.

CHAPTER SYNOPSES (CHAPTERS 2–12)

Chapter 2: Success Starts Upfront: Describe the Problem Accurately

The program management team plays a critical role during the program frame-up phase by ensuring that there is due diligence around business problem definition and alignment among the stakeholders on the problem the business transformation program will solve. The common understanding and agreement of the problem statement among the stakeholders creates an invaluable platform that the program management team can further build upon as they successfully drive the program forward.

The following topics are described in this chapter with the help of supportive illustrations, including a real world case study:

- Program management operating model
- Program management tool kit
- Environment scanning technique: Overview, objective, and approach
- Description of business problem
- Voice-of-customer technique: Overview, objective, and approach
- Definition of business outcome
- Definition of benefits realization
- Program charter
- Case study: Software licensing transformation program

Chapter 3: Articulate the Program Vision and Objectives

As the program management team drives forward the planning and execution of programs, all of the program stakeholders need to be cognizant of the program vision, program strategy, and program objectives. Periodic reinforcement of the program vision and of how achievement of the program objectives will benefit everyone is one of the most important and critical success factors for a transformation program.

The following topics are described in this chapter with the help of supportive illustrations, including a real world case study:

- Formulation of business strategy
- Business transformation drivers
- Implementation of business strategy
- Comparing program management life cycle to PMI's *Standard for Program Management*
- Strategic alignment technique: Overview, objective, and approach
- Business performance calibration technique: Overview, objective, approach, and critical success factors
- Strategic imperatives architect programs
- Benefits realization strategy
- Management of benefits realization life cycle
- Case study: Transformation program to redesign process and technology

Chapter 4: Secure Cross Functional Executive Sponsorship

The sponsorship team needs to own the transformation program and be visibly involved. The drivers for the business transformation need to be reinforced by the sponsors. A cross functional and multilevel sponsorship model is needed to authorize and legitimize the program. By addressing the program barriers, the sponsorship team positions the program team to realize the desired business outcomes expected of the transformation program.

The following topics are described in this chapter with the help of supportive illustrations, including a real world case study:

- Sponsorship of business transformation program
- Multilevel program sponsorship model
- Program value justification technique: Overview, objective, approach, and helpful hints
- Upward management of program sponsors
- Sponsorship of program outcome delivery and benefits realization
- Case study: Program to transform procurement function via outsourcing

Chapter 5: Develop and Implement a Governance Model

For a transformation program to succeed, it is imperative to develop and implement the governance practices sooner rather than later. A program-governance model is a combination of governing bodies, strategic control and oversight functions, and cohesive policies that defines the consistent management of the program throughout the program life cycle. A lack of robust governance practices poses a substantial risk to realizing the desired business outcomes targeted by the transformation program.

The following topics are described in this chapter with the help of supportive illustrations, including a real world case study:

- Governance modeling technique: Overview, objective, and approach
- Program communication and escalation protocol
- Program governance: Bodies, responsibilities, and rhythm
- Program accountability
- Governance policy design technique: Overview, objective, and approach
- Program governance: Decision making framework and change control management

- Governance of benefits realization
- Case study: Transformation program for postmerger integration

Chapter 6: Define Success, Outcomes, and Key Value Indicators

The importance of driving adequate clarity upfront on what constitutes the success of a transformation program is high. The elaboration of expected outcomes and timing the delivery of those outcomes is a critical success factor. Program management has to constantly manage stakeholder expectations and validate that there are no deviations from the original definition of success. The activities associated with management of the business outcome life cycle have to be embedded in the integrated transformation program plan.

The following topics are described in this chapter with the help of supportive illustrations, including a real world case study:

- Performance improvement measurement technique: Overview, objective, approach, and helpful tips
- Positioning transformation program to deliver business outcomes
- Business-outcome modeling technique: Overview, objective, and approach
- Case study: Selection program for enterprise system

Chapter 7: Invest in Planning and Creating an Integrated Approach

An end-to-end and integrated program management approach will result in the accomplishment of strategic objectives and sustainment of defined business outcomes. The program architecture establishes the bridge between business strategy and the implementation of that strategy, which provides the key input to the program management life cycle. The first two phases of the program management life cycle are essentially planning phases, and the culmination of those then results in fruition of a cohesive, integrated program plan. Organization change management (OCM) is a blueprint for integrating and accounting for the "people" elements of the transformation program. Program management and organization change management are two sides of the same coin, with the former focusing on the "hard or technical or structural" aspects and the latter on the "soft or cultural or behavioral" aspects, and both are critical to enable and embrace the strategic business change.

The following topics are described in this chapter with the help of supportive illustrations, including a real world case study:

- Program architecture technique: Overview, objective, and approach
- Program architecture drives program management life cycle
- Organization change management technique: Overview, objective, approach, and drivers
- Organization change readiness
- Transformation program planning technique: Overview, objective, and approach
- Benefits realization planning
- Case study: Business transformation program to launch a new product

Chapter 8: Drive Strong Partnership and Stakeholder Engagement

For a transformation program to successfully deliver and sustain the change, stakeholder engagement at all levels is necessary. A complex program is usually driving strategic business and technological change within multiple functional groups. Typically, a vast number of business and technical stakeholders in multiple geographic locations at varying organizational levels are impacted by the program. The full support of as many stakeholders as possible to attain and sustain the desired future state is essential. The program is considered successful by stakeholders if it delivers the expected business outcomes and benefits. The program leaders and core program team members need to proactively and constantly partner and collaborate with the identified stakeholders. A tailored engagement strategy and an engagement plan need to be crafted for each category of stakeholders.

The following topics are described in this chapter with the help of supportive illustrations, including a real world case study:

- Stakeholder expectation management technique: Overview, objective, and approach
- Stakeholder assessment
- Stakeholder engagement strategy

- Stakeholder engagement: Planning, executing, and monitoring
- Stakeholder engagement through communications vehicle
- Stakeholder engagement through training vehicle
- Stakeholder engagement through coaching vehicle
- Business outcome delivery enabler
- Case study: Transformation of policy servicing at enterprise level

Chapter 9: Provide Leadership Across All Levels

The complexity of business transformation programs requires cross functional, cross-dimensional, and cross-project leadership to realize the program vision and sustain business outcomes. Effective leadership at multiple levels is needed to increase the acceptance rate of the transformational change brought about by a program. The office of business transformation provides an integrated leadership model that facilitates getting the right quantum of leadership to all the dimensions of program management.

The following topics are described in this chapter with the help of supportive illustrations, including a real world case study:

- Multilevel program leadership model
- Multidimensional program leadership model
- Program leadership and risk mitigation
- Transformation program leadership: Planning and delivery stages
- Leadership of program management processes
- Program leadership versus program management
- Balancing program leadership and program management
- Program leadership through office of business transformation
- Drivers for the office of business transformation
- Significance of leadership in program communications
- Benefits realization leadership
- Case study: Business transformation initiative on privacy and compliance

Chapter 10: Monitor Aggressively and Have Contingencies

Institutionalization of program controls is essential for efficient operational execution of the business transformation program and to position the program for success. The "monitor program delivery" process ensures that program progression is in line with the integrated program plan. The program monitoring process drives additional value into the business-transformation program by creating opportunities for continuous improvements. The tested contingency plans mitigate the significant risks encountered by the transformation.

The following topics are described in this chapter with the help of supportive illustrations, including a real world case study:

- Monitoring status of transformation program
- Program monitoring: Key criteria
- Status of transformation program dashboard
- Tracking and reporting project performance
- Monitoring transformation program: Milestones, financials, and issues and risks, and change requests
- Enabling continuous improvement
- Contingency strategy and contingency planning
- Monitoring benefits realization
- Case study: Program to manage transformation of business processes

Chapter 11: Create and Implement an Operations Transition Plan

The timely roll-out of an end-to-end operations transition framework will position the designated operations functions to move forward the charter of the business transformation program once it ceases to exist. The readiness of the operations team is facilitated by the development and execution of the operations plan, training plan, and knowledge transfer plan. Embedding continuous improvement practices and lessons learned exercises creates an operations environment that will sustain the delivery of expected business outcomes.

The following topics are described in this chapter with the help of supportive illustrations, including a real-world case study:

- Operations transition framework
- Transition model development
- Training the operations team
- Transition model execution
- Value enhancement analysis technique: Overview, objective, and approach

- Transition model operationalization
- Walkthrough technique: Overview, objective, and approach
- Leading lessons learned from walkthroughs
- Business outcome delivery and sustainment
- Case study: Strategic initiative to get and stay compliant with government regulations

Chapter 12: Executive Summary

For business transformation programs to succeed in today's ever-changing market environment, enterprises must give renewed emphasis to the tenets of program management. The program management discipline integrates and aligns the six critical dimensions (strategy, people, process, technology, structure, and measurement) needed to transform a business through successful completion of a transformation program. For organizations seeking a competitive edge, program management is the "secret sauce" in achieving transformation program objectives, realizing program outcomes, delivering business results, and setting the stage for ongoing benefits realization.

The following topics are described in this chapter with the help of supportive illustrations:

- Road map for transformation (or strategic initiative) success
- Summaries of Chapters 2–11
- Revisit of main points
  - Program architecture: Bridge to implementing business strategy
  - Business outcome and benefits realization life cycle management
  - Program management life cycle
  - Program management office and office of business transformation
- Key takeaways on program management

TRANSFORMATION PROGRAM MANAGEMENT CLOSES THE BUSINESS OUTCOME GAP

Enterprises everywhere are experiencing the *business outcome gap*, which can be defined as the difference between desired business outcomes and realized business outcomes. Desired outcomes are changing in response to rapidly evolving stakeholder needs, whether the stakeholders are employees, customers, or shareholders. While realized outcomes may be improving, for most enterprises the increase in desired outcomes is far outstripping those realized. Not only do enterprises see a business outcome gap, but they see it widening. To close the business outcome gap, enterprises are increasingly embarking on transformation initiatives or strategic initiatives. These initiatives are launched by senior management within an organization to transform the business from the current to the future state and, in doing so, to close the widening business outcome gap.

Enterprises need to understand and address the six dimensions of program management (strategy, people, process, technology, structure, and measurement) in any transformation initiative they attempt. The multidimensional and integrative framework of program management enables enterprises to initiate, plan, execute, monitor, deliver, and sustain the benefits expected by the stakeholders of the transformation initiative. Program management ensures that the planned tangible and intangible benefits are realized in the designated time frame and that the realization of benefits results in a closure of the business outcome gap.

TRANSFORMATION PROGRAM REALIZES BUSINESS BENEFITS

A business transformation program is designed to accomplish strategic business objectives, which tend to be high level. Typically, the program charter will expand on each strategic business objective and derive the desired business outcomes. In other words, the program team decomposes the business objectives into desired business outcomes. The desired business outcomes are further decomposed into specific business benefits (or business results) expected by stakeholders. Hence, the business outcomes realized by the program are the business benefits realized by the program. The expected benefits (or results) to be realized by the program could be tangible, intangible, or both. The socialization and sign-off of the transformation program charter by all stakeholders in the early stages of the program management life cycle ensure that there is buy-in on the business objectives, business outcomes, and business benefits to be delivered by the transformation program.

A program is designed to realize tangible and intangible business benefits, and these benefits are sustained by the operation's function upon formal closure of the program. The program management team works closely with the stakeholders in devising the benefits realization plan and secures their sign-off on the plan. In addition, the program manager gets buy-in and approval of the benefits realization plan from the executive sponsor. As the program management life cycle plans and executes the transformation program, the tangible and intangible benefits specified in the approved benefits realization plan are delivered. The technique of business-outcome modeling is covered in Chapter 6, and the implementation of that technique adds rigor to the steps taken for a program to realize the planned or expected benefits. The business outcome realization plan is a component of the transformation program plan, which is described in Chapter 7 of the book.

CLARIFICATION ON BENEFITS REALIZATION TERMINOLOGY

The ultimate goal of the business transformation program is to realize the vision and strategy within the planned time frame and with the estimated resources. The realization of program vision means the realization of business objectives, business outcomes, and business benefits. Business benefits can be categorized as either tangible or intangible. The transformation program is designed to realize the intended business benefits in the desired future state and then sustain those benefits. The extent of success of a program is determined by tracking the business benefits in the attained future state. An example of a business objective for a widget manufacturing organization might be to "Be the market leader in the United States."

An example of a business outcome supporting this business objective might be to "Be the number 1 in total volume and total dollar sales of the widget in three of the four geographic regions within the United States in which the organization operates." An example of a tangible business benefit supporting this business outcome might be improved top-line financials. An example of an intangible business benefit supporting this business outcome might be to "Be the most admired brand." In this book, the terms *business objective, business outcome,* and *business benefit* are used in this context.

BUSINESS TRANSFORMATION AND PROGRAM MANAGEMENT LIFE CYCLE

Enterprises are systemic and complex in nature, and many are global. Strategic initiatives (or transformation programs) undertaken by enterprises are complex and drive significant business change across autonomous business divisions, functions, and geographic locations. A typical strategic initiative at a large enterprise is a complex program comprising multiple related projects. Management of strategic or transformation initiatives is called *program management.*

Program management enables and sustains business transformation by articulating vision, developing an integrated transformation program plan, driving the plan, removing execution barriers, delivering planned business outcomes, and realizing business benefits. Though program management is not a silver bullet, it can play a key role in realizing the business benefits (or business outcomes), as it can effectively account for and address all six dimensions of the program management life cycle: strategy, people, process, technology, structure, and measurement. True, sustainable business outcome realization is only achievable by recognizing, analyzing, and addressing the six dimensions.

The program management function within an organization can facilitate and own the responsibility for executing a transformation program. The discipline and rigor in planning and executing the transformation (or strategic) initiatives can be ensured and facilitated through the program management life cycle. In the context of program management, *life cycle* refers to program start to program finish. The program management life cycle (Figure 1.3) comprises four phases:

Phase 1: Set the stage

The primary objective of this "set the stage" phase is to validate the program mission, vision, and strategy and define the problem. Program objectives, desired business outcomes, and a high-level program plan is developed. This goal is accomplished by completing the following two high-level processes:

- Formulate a program strategy
- Develop a program road map

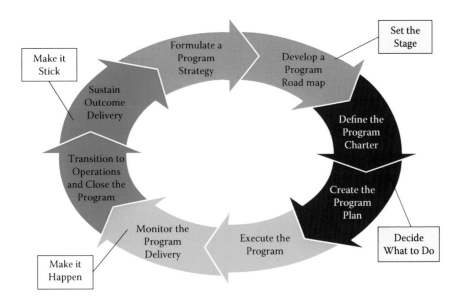

FIGURE 1.3

Program management life cycle.

Phase 2: Decide what to do

In the "decide what to do" phase, the definition of success, problem validation, and program scope determination is completed. In addition, stakeholders are identified, an integrated detailed program plan is created, and program plan is baselined. The "decide what to do" phase entails the below processes:

- Define the program charter
- Create the program plan

Phase 3: Make it happen

The work that needs to get done as part of the program happens in this "make it happen" phase. As the program plan is executed, the program monitoring function starts. This real time monitoring enables timely intervention, facilitates course correction, and validates program output is in line with the plan. The two high-level processes that accomplish the goals of "make it happen" phase:

- Execute the program
- Monitor the program delivery

Phase 4: Make it stick

The "make it stick" phase entails the various steps associated with the transition of program to the operations function including transfer of knowledge. Realized program business outcomes are compared against expected program business outcomes and lessons learned analysis is accomplished. The following two processes constitute this "make it stick" phase:

- Transition to operations and close the program
- Sustain outcome delivery

SUMMARY

As organizations become more global and their transformation initiatives get more complex, effective program management will be critical to achieving the strategic objectives. In light of the disruptive changes that the transformation initiatives are driving, the complexity is high and the business change is significant. Enterprises need a methodical approach to successfully execute a transformation initiative, and a road map for the same was presented in this chapter. In addition, enterprises need a comprehensive framework for a transformation program to deliver the desired business outcomes and realize the planned business benefits.

The program management framework is the glue that brings strategy, people, process, technology, structure, and measurement dimensions together to enable the realization of benefits expected from a transformation initiative.

This chapter identified fifteen program management techniques that can be leveraged during the course of a transformation program to successfully transform a business. The transformation program road map and the program management techniques will be further expanded in the remainder of the book.

Program management can be instrumental in closing the gap between the desired and realized business outcomes. The program management life cycle facilitates the end-to-end process of planning through the eventual execution of the business transformation program. For organizations seeking a competitive edge, program management is the "secret sauce" in achieving transformation program objectives, realizing program outcomes, delivering promised business results, and setting the stage for ongoing realization of benefits.

REFERENCE

Economist Intelligence Unit. 2013. *Why good strategies fail: Lessons for the C-Suite.* London: Economist Intelligence Unit Limited.
Subramanian, Satish P. *The Steps to Program Success.* PM Network, September 2012.

2

Success Starts Upfront: Describe the Problem Accurately

A business transformation program can fail for many reasons, and inadequate definition of the business problem is certainly a leading one. Many times, there is a lack of upfront rigor in elaborating, defining, and validating the business problem. Enterprises are unique; their problems are unique; and hence each solution to these problems must be unique. More emphasis should be placed on the thoughtful definition of the problem to be solved and on the careful selection of appropriate methods to solve it. There is a tendency for enterprises to look for historical or easy solutions. A particular method of successfully solving a strategic business problem in one company may not work equally well for another company with the same problem. A business transformation program chartered to address an incorrect problem is inherently doomed to fail, so a program manager must determine the exact strategic business problem from the outset.

The following topics are described in this chapter with the help of supportive illustrations, including a real world case study:

- Program management operating model
- Program management tool kit
- Environment scanning technique: Overview, objective, and approach
- Description of business problem
- Voice-of-customer technique: Overview, objective, and approach
- Definition of business outcome
- Definition of benefits realization
- Program charter
- Case study: Software licensing transformation program

PROGRAM MANAGEMENT OPERATING MODEL

A transformation program has to effectively solve today's complex business problems in a constantly changing and competitive business environment. Program management integrates multiple disciplines and, therefore, is uniquely positioned to solve the problems or capitalize on the opportunities that business transformation initiatives undertake. Chapter 1 introduced the high level, six dimensional program management framework. The transformation program team management implements the program management framework through the program management operating model.

This operating model provides a unique platform to transform the business by integrating the six dimensions with the seven stages of the solution life cycle. This solution life cycle is captured in the horizontal axis, and the six dimensions (strategy, people, process, technology, structure, and measurement) are reflected on the vertical axis. The construct of a program management operating model enables the program management team to factor the six dimensions across the solution life cycle, which moves the organization from the current to the future state. The application of the program management operating model positions the business transformation initiative to be successful by delivering desired business outcomes, realizing expected business benefits, and producing measurable business results.

Solution Life Cycle

The solution life cycle is an end-to-end problem solving (or opportunity capitalization) cycle starting with definition of a business problem and ending with sustainment of a benefits outcome. The solution life cycle comprises the following seven stages, and these are the columns in Figure 2.1.

- Definition of business problem
- Determination of objectives and outcomes
- Program design and planning
- Stakeholder identification and engagement
- Program execution and tracking
- Realization of program outcomes
- Sustainment of outcomes and results

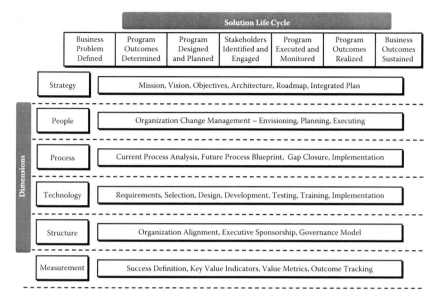

FIGURE 2.1
Operating model for program management.

Program Management Dimensions

The high level building blocks within each of the six dimensions are highlighted in the rows in Figure 2.1. As an example, the key building blocks constituting the "process" dimension are the current process, the future process, gap closure, and implementation of the future process. Similarly, the key building blocks for the "measurement" dimension include success definition, key value indicators, value metrics, and outcome tracking. As the program management team develops the program plan for executing the business transformation, the team has to ensure that all of the six dimensions are being taken into consideration for each stage of the solution life cycle.

For example, during the "business problem definition" stage of the solution life cycle, the analysis and engagement with program stakeholders could lead the program manager to conclude not only that the technology solution is not meeting the end user needs, but also that the end-user personnel have not been trained. In this scenario, the solution that does not meet the end-user needs reflects the "technology" dimension, and the lack of training of end-user personnel reflects the "people" dimension. Based on the insights gathered in this case, in the program plan, the program manager has to detail all the activities associated with both of these dimensions.

PROGRAM MANAGEMENT TOOL KIT

The program manager takes advantage of the numerous tools and techniques from the program management tool kit. The program management maturity level of an enterprise dictates the comprehensiveness of the tool kit and how it has been tailored over the years to successfully handle a wide range of programs. Many organizations use a *program management office* or a *program management center of excellence* to develop, pilot, refine, and maintain such a tool kit. Some programs, especially ones that are large and complex, may have their own program management office (PMO) to support the program and use and tailor the tool kit.

The following discussion highlights two techniques that enterprises and program managers have successfully applied to identify, define, detail, and validate business problems. Each of the fifteen techniques covered in this book includes an overview ("what"), objective ("why"), and an application approach ("how"). Though a program manager need not be a subject matter expert in all of the techniques applied to drive a business transformation initiative to success, it is important for the program manager to have a good understanding of these techniques. A program manager can recommend, plan, facilitate, and manage the implementation of all techniques and be an expert in some of the techniques in the program management tool kit.

ENVIRONMENT SCANNING TECHNIQUE

Overview

An environment scan is the identification and structured analysis of strengths, weaknesses, opportunities, and threats in the context of organizational strategy. Strengths and weaknesses are characteristics of the enterprise or business process—its operational assets and liabilities. Most strengths and weaknesses exist in a tangible sense. For example, an asset such as an extensive distribution network is a strength regardless of environmental factors. Opportunities and threats include internal or external environmental factors or trends that can positively or negatively affect an enterprise, depending on where its strengths and weaknesses lie. Environment scanning forms the premise for business problem identification, definition, and validation and provides a vehicle for an enterprise to effectively and accurately frame a problem.

Objective

The primary purpose of environment scanning is to assist the enterprise in developing a strategy that takes advantage of opportunities and overcomes or circumvents threats. This scanning analysis is not always easy, as it can sometimes be difficult to identify which environmental changes constitute an opportunity and which constitute a threat. The process of environment scanning maps opportunities and threats to both strategies and environmental factors. It provides an at-a-glance picture of factors that affect strategies and the environment. A structured analysis is required to identify how an enterprise's current strategy, strengths, and weaknesses determine the opportunities and threats from its business environment. Determining weaknesses assists in problem identification and definition. Similarly, a close examination of the threats provides insight on what could become a significant problem.

Approach

The process for performing environment scanning involves the following steps:

- Determine the level of analysis
- Identify the key strengths and weaknesses
- Analyze the strategy against the strengths and weaknesses
- Identify key changes in the environment
- Map environmental influences and changes
- Analyze environmental influences and changes
- Identify opportunities and threats

A detailed description of each of the above steps constituting the environment scanning technique follows as it will facilitate its effective application by the business transformation program core team in order to produce the needed deliverables.

Step 1. Determine the level of analysis: An analysis can be performed at a variety of levels: enterprise wide, business unit, group of functional areas, single function, or specific business process. At each level, the same steps apply, but the difference is the extent of detail and the environmental factors involved. The level must remain the same throughout the analysis.

Step 2. Identify the key strengths and weaknesses: The next step is to analyze the resources at the enterprise, business unit, function, or business-process level to provide an understanding of strategic capabilities. These capabilities provide the basis for the enterprise's competitive position and for its ability to fulfill the roles expected by customers and other stakeholders. The objective of this step is to list specific strengths and weaknesses. This list is not simply a quantitative inventory of assets and liabilities, but is an assessment of capabilities in relation to the strategic purpose, or mission, of the enterprise or function. If an enterprise decides to further build on particular strengths to increase the business performance gap with its closest competitors, the problem definition statement morphs into strategic business objective statements.

Step 3. Analyze the strategy against the strengths and weaknesses: Analyzing the enterprise's strategies in relation to its strengths and weaknesses is the next step. If the analysis is being conducted at the business process level, analyze the associated key value indicators (KVIs) in relation to the strengths and weaknesses. KVI is an operational performance measure used to monitor the success of the transformation program in realizing the desired business outcomes. KVI assists with measuring attainment of business objectives, business outcomes, and business benefits. As programs drive business initiatives forward with the help of process focused cross functional teams, operationally based performance measures become key in measuring the value delivered by the program.

A *Harvard Business Review* article (Meyer 1994) has stressed the application of process based measures in a performance-measurement system designed to gauge the value in delivering a service or product to customers. KVIs are extensively covered in Chapter 6. Note that these are not the same as key performance indicators (KPIs), which measure program work performance and progression against the program plan. The analysis of the strategy by the core program team, subject matter experts, and other stakeholders ascertains whether the current strategy capitalizes on strengths or addresses weaknesses.

Step 4. Identify key changes in the environment: The objective of this step is to develop a short list of environmental changes or influences that most affect at the enterprise, business unit, function, or business process level. There are a number of different types of changes and influences to consider. Influences in the economic, technological, political, and sociocultural environments should be examined. Examination these influences is necessary, as they impact the organization's vision, strategy, business objectives, and desired business outcomes.

For example, if the manufacturing operations of a global company are concentrated in a geographic region where political instability and uncertainty jeopardize the prevailing labor laws, the environment scanning would be used to assess the situation. At an industry level, competitive rivalries, buyer and supplier relationships, and the threat of substitute products and services should be considered. For example, if a substantial portion of a software company's revenue comes from the sales of on-site based software while competitors are gaining market share with cloud based software, the software company cannot afford to ignore the threat.

In this step, it is critical to keep in mind the level of analysis being performed. The environmental factors that are relevant to the global enterprise are significantly different from those relevant to the business process. The influencing factor categories (economic, technological, political, and sociocultural) remain the same, but the specific factors will be different and will tend to be more specific with greater granularity of the environment scanning. In other words, an analysis of a business process would be more affected by a specific technological innovation than a general technological trend. To extend the example presented here, the software company that is currently losing market share would identify the environmental change in the software licensing business process. Two effective methods for compiling the list of environmental factors include (a) brainstorming sessions involving the program team and other affected stakeholders and (b) creating a process for each stakeholder

Environment Scanning Analysis						
Key Strengths and Weaknesses	Technology Developments	Consumer Tastes	Industry Overcapacity	Globalization	Aging Population	Talent Availability
Strengths						
Innovation	+	+	−	O	O	−
Financial Capital Availability	+	+	+	+	+	+
Brand Reputation	O	−	O	−	+	−
Information Systems	+	O	−	+	−	O
Weaknesses						
Product Portfolio	−	+	−	−	O	−
Scalability	O	O	+	−	+	−
Supply Chain	+	O	−	+	O	−
Manufacturing	−	−	+	−	+	−

FIGURE 2.2

Environment scanning analysis: identifies opportunities and threats.

to independently develop the list followed by a team sharing session to reach group consensus.

Step 5. Map environmental influences to the strengths and weaknesses: The next step is to develop a matrix to map the key issues in the environment against the strengths and weaknesses identified in the previous step. In this way, the enterprise can establish the relevance and significance of the environmental factors in terms of the strengths and weaknesses of the organization. Figure 2.2 provides an illustration of the matrix, which is one of the key inputs that could be used in the business problem identification and definition process.

Step 6. Analyze environmental influences and changes: The next step in performing an environment scanning analysis is scoring each of the entries in the matrix in Figure 2.2. The goal of this step is to determine whether internal and/or external environmental factors and trends provide opportunities or threats for the enterprise given its current strategies, strengths, and weaknesses. While the enterprise may have its own weighted scoring systems, a simple method involves using pluses, minuses, and zeros. This scaling is performed as follows:

- +: used to indicate a benefit or opportunity for the enterprise. This score could be either a strength that will enable the enterprise to take advantage of an opportunity or a weakness that would be offset by an environmental change.

- −: used to indicate adverse effects on the enterprise in the form of an organizational strength being reduced by an environmental change or a weakness preventing the enterprise from handling an environmental change.
- o: used to imply neutrality, i.e., the environmental trend has no effect on a given organizational strength or weakness.

The result of the scaling is a clearer insight of the extent to which environmental changes and influences provide opportunities or threats given the enterprise's current strategies and capabilities. This deeper level of analysis assists in validating the accuracy of the defined problem, which is another key factor in supporting the business rationale for initiating a transformation program. A sample output of the analysis is highlighted in Figure 2.2.

Step 7. Identify opportunities and threats: The final step in environmental scanning identifies the opportunities and threats created by the interaction among the enterprise's strengths, weaknesses, and environmental influences. If an environmental trend provides an opportunity based on the strengths, any strategies relying on those strengths should be emphasized to take advantage of the opportunity. Similarly, if an environmental trend poses a threat, there may be a cause for a new strategy to negate that threat. If a trend negates the strengths of the enterprise, any strategies reliant on those strengths should be modified. An enterprise might view the lack of any initiative to capitalize on known opportunities as a problem and decide to initiate a transformation program to sustain the competitive advantage in that area.

DESCRIPTION OF BUSINESS PROBLEM

Earlier in this chapter, the discussion of a program management operating model emphasized the need for an accurate and comprehensive definition of the problem. The first two stages of the solution life cycle are (a) the definition of the business problem and (b) determination of the objectives and outcomes. All six of the program management dimensions (strategy,

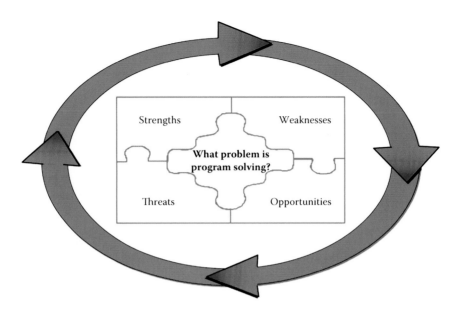

FIGURE 2.3
Definition of business problem: leveraging environment scanning analysis.

people, process, technology, structure, and measurement) have to be taken into consideration as part of the solution life cycle. The research, deliverables, and information gathered from an environment scanning analysis exercise should be leveraged to define, describe, and validate the business problem (Figure 2.3). The transformation program management team effectively engages with stakeholders to ensure the alignment of the problems that need to be solved first. Agreement and clarity on the question of "What problem will the program solve?" will provide a solid footing to describe the program objectives and outcomes.

VOICE-OF-CUSTOMER TECHNIQUE

Overview

The second technique from the program management tool kit that can be utilized to uncover, define, and validate the business problem is the Voice-of-Customer (VoC). The importance of factoring in an "outside in" perspective is significant during the problem identification and elaboration processes. During this early discovery stage of the transformation initiative, as the program is being established, the program manager can bring

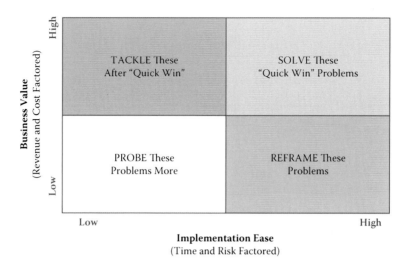

FIGURE 2.4
Solving the right problem: applying the voice of the customer.

in this external view through a Voice-of-Customer study. The VoC is a technique used to identify and validate the customers for the transformation program and to capture customer insight on their pain points, what they value the most, expected business outcomes, and points of view on how well the program will address their needs. In *Satisfaction: How Every Great Company Listens to the Voice of the Customer* (Denove and Power 2007), the authors share fascinating stories of companies that don't just talk the talk, but walk the walk every day, and of other companies that ignored the voice of the customer, with dire consequences (Figure 2.4).

Objective

The application of the VoC technique:

- Identifies problems and assists in accurately framing them
- Enables an objective view of what is important to customers
- Segments customers and diagnoses the value criteria for each segment
- Facilitates the identification and prioritization of what is important to customers in influencing their decision making process
- Measures customer satisfaction across various performance criteria
- Identifies where improvements in the business are needed
- Assists in the identification of key value indicators

The customer's view of "value" is independent of the actual costs incurred by the enterprise, business unit, function, or business process that provides the product or service. In this respect, it is necessary to discern the specific criteria by which customers see value in the products and services they purchase. The VoC thus provides additional following invaluable inputs into the program:

- Allows the enterprise to focus strategies and processes around customer needs
- Identifies strengths and weaknesses according to the customers
- Compares current performance to competitors' performance on key customer values
- Provides intelligence about competitors
- Helps in determining which processes and technology systems are value added and how to innovate them to enhance customer experience

Approach

When conducting a VoC analysis, follow these steps:

- Identify customers
- Determine needs of customers
- Construct "business value–implementation ease" matrix
- Reread the marketplace

Each of the above steps constituting the VoC technique is elaborated below as it will facilitate an effective application of this technique by the business transformation program team in order to generate the necessary outputs.

Step 1. Identify customers: A best practice is to segment customers into appropriate groups that demonstrate similarities. The segmentation criteria utilized will vary, but essentially items such as sex, age, geographical location, product type, etc., will be considered. The criteria can be enhanced by answering the following:
- Who is the recipient of the output from the activity and process?
- How do the consumers' needs and wants affect the activities within the process?
- What do they expect to receive?

- How is the output expected to be utilized?
- If the output does not meet the initial customers' expectations, what is the impact?

The customer identification step assists with classifying the different types of customers that exist. Typically, the customer would fall under one of the following four categories:

- **External customers:** These are customers outside the organization that receive the end product or service.
- **Consumers:** For some organizations, the consumer and the external customer are one and the same, whereas for the others the consumer is an indirect external customer.
- **Indirect internal customers:** These are customers that are not within the functional area but who receive output from the activity or process.
- **Direct internal customers:** These are the customers within the functional area who directly receive the output from an activity or a process as a whole.

In business-transformation programs, a focus on external customers is necessary because they are the most important in determining enduring competitive success.

Step 2. Determine the needs of the customers: Customer needs can be gathered through customer focus groups, in-person interviews, surveys, telephone interviews, questionnaires, and meetings. Either a high-touch method (e.g., focus group) or a low-touch method (e.g., web-based surveys) or a combination thereof can be used in determining the needs of the customers in each category. The goal is to collect high quality information from a wide range and large number of customers and analyze that information efficiently. The information gathered will be used to determine customer value criteria. It is important to properly use a well-developed questionnaire. Study the actual activities of consumers using the product or service. Where are they? Where do they use it? Who is influencing them? It is important to make a detailed list of the different needs that are being met.

Typical examples of customer value criteria that will be revealed are:

- Product and service feature richness
- Rapid availability
- Flexible delivery capabilities
- Price

- Post-sales service and support
- Product or service brand image

Step 3. Construct a "business value–implementation ease" matrix:
Analyze the customer needs and highlight the differences between customer values and enterprise values. Regardless of how long a company has been in a particular market, running optimized operations, or monitoring current market intelligence, the VoC findings help in understanding the customer values and business problem areas. The program management team can facilitate the creation of a "business value–implementation ease" matrix, which is a result of the problems getting analyzed by severity, business impact, complexity, time, resource needs, business risk, implementation feasibility, and cost. The vertical Y axis represents the business value to the organization in putting the business problem behind by factoring in revenue and cost. The horizontal X axis is an indicator of how easily the business problem can be addressed. The resulting matrix provides insights that can drive the direction to be taken to effectively solve the problem and/or exploit an opportunity.

The "business value–implementation ease" matrix is an invaluable tool in prioritizing the identified problem(s) that need to be addressed to meet the needs of the customers. The matrix aids in making key decisions, justifying the decisions to senior management, and determining next steps. The matrix highlights four potential options:

- Solving those problems that would result in a "quick win"
- Tackling the next set of problems which will also deliver high business value
- Reframing other problems with the intent of deriving higher business value
- Probing deeper into the leftover problems to get more clarity

Thus, the matrix guides the program team in prioritizing the problem solving initiatives within the program road map, which is typically a multiyear, high level plan in the case of a transformation program. Figure 2.4 displays the "business value–implementation ease" matrix.

Step 4. Reread the marketplace: After the initial customer driven needs, pain points, and innovation opportunities have been captured, the enterprise, business unit, business function, or business process team must go back and resurvey the marketplace to see what innovations they would consider to be valuable, e.g., "Does the market want what the enterprise believes it can provide?" This second round of VoC and market research validates whether the strategic direction of the transformation program is in line with the confirmed needs. The core program management team collaborates with the subject matter experts in the various functional organizations in completing the VoC study and gets their recommendations to address customer pain points. The program management team then reviews the completed VoC analysis with the transformation program stakeholders and incorporates the work to be completed by the program for a delivered future state that will overcome the pain points.

DEFINITION OF BUSINESS OUTCOME

The success of a transformation program is primarily gauged by the delivery of business outcomes. Figure 2.5 illustrates the approach to be taken in the front end of the program life cycle to design, influence, and plan the business outcomes. A revisit of the business and technology strategy

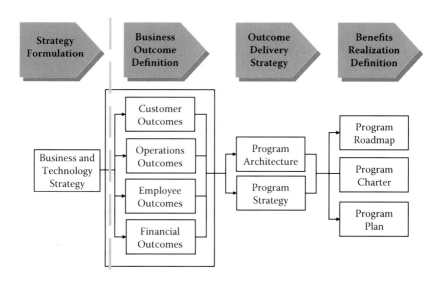

FIGURE 2.5
Positioning the program to deliver business outcomes.

is done to identify the business-outcome components, which would vary by the nature of program. Typically, at the highest level, the strategic outcome components would fall in the areas of customers, employee operations, and finance. The program architecture and program strategy stages will articulate these business outcomes to the next level.

The following considerations are necessary to effectively define the business outcomes:

- In developing *customer related outcomes*, it is important to know if the customers can be segmented by different characteristics. If distinct customer segments exist, relevant business outcomes would need to be selected for each segment. Examples of potential desired business outcomes include growth in market share, higher customer satisfaction, on time delivery, etc.
- *Operational outcomes* developed should reflect the cross functional or cyclical nature of how the work is performed within an organization. For instance, lower cycle time, higher throughput, cost reduction, etc., would be sample business outcomes.
- *Employee related desired business outcomes* must foster employee growth and happiness, which consequently generate improvement in the other outcome components. Some examples here are an empowered workforce, revenue per employee, new product ideas, etc.
- *Financial outcomes* result in shareholder value delivery and tangible performance improvement to run the business, e.g., revenue growth, higher profitability, improved cash flows, etc.

In the program architecture and program strategy stages, the business outcome delivery strategy is formulated. The outcome delivery strategy facilitates alignment of the defined business outcome components. The correlation amongst customers, operations, employees, and financial outcomes is factored in crafting the outcome delivery strategy.

DEFINITION OF BENEFITS REALIZATION

The development of a program road map, program charter, and program plan supports implementation of the business outcome delivery strategy.

The program manager works with the key stakeholders in determining the tangible and intangible business benefits the transformation program is chartered to deliver and sustain. For each identified and agreed business outcome component, the elaboration of tangible and intangible business benefits to be realized by the program needs to be completed. For example, as part of defining the financial outcome component, the tangible top line and bottom line financial benefits (or results) the program is expected to deliver upon completion could be specified. The realization of the defined business benefits can be gauged with the help of the key value indicators (KVIs). The KVIs are high level, and the underlying value metrics aid in confirming the realization of the business benefit.

Chapter 6 covers the benefits realization measurement architecture, KVIs, and value metrics in more depth. The definition of benefits to be realized by the program entails not just the description of the tangible and intangible benefits, but also the timing of the realization of those benefits. The program management team utilizes the program road map and program plan to drive the transformation program forward and realize the defined benefits in the specified time frames. Although the section in Chapter 7 on benefits realization planning explicates the timing of the benefits realization in depth, typically the strategy is to realize some benefits as the program is underway, while others are delivered at the end of the program. The upfront identification and definition of the tangible and intangible benefits to be realized and sustained by the business transformation program is crucial.

PROGRAM CHARTER

The program charter aids in detailing the problem statement, articulating the definition of success, setting scope boundaries, and securing stakeholder buy-in on outcomes. In addition, the charter provides the program manager with the authority to assign resources to the business transformation program. The program charter, a core deliverable in the early stages of the program life cycle, forms the basis for many of the other core deliverables of program management. This tangible artifact can assume different names within different enterprises based on the program management terminology in place, e.g., program definition document, program description document. Similarly, the level of detail and subject matter topics included in this deliverable depend on the program management methodology and program management maturity level of the enterprise.

At a minimum, this deliverable has to clearly articulate the business problem, the program objectives, what constitutes program success, program scope, problem solution approach, a high-level timeline, and budget estimates. The best practice approach is for the program sponsor to prepare the program charter with input from the program manager. An alternative route is for the program manager to take the lead in preparing the charter with high involvement of the program sponsor. The instructions for developing a program charter are outlined in detail in *Implementing Program Management: Templates and Forms Aligned with the Standard for Program Management* (Levin and Green 2013). The program manager ensures that the program charter has been signed off by all the key stakeholders.

Figure 2.6 illustrates the composition of a typical program charter for a complex business transformation program. The sample charter comprises twelve building blocks. The first four blocks describe the problem the program has been chartered to address, the objectives to be accomplished, the capabilities to be developed, and the benefits the program will

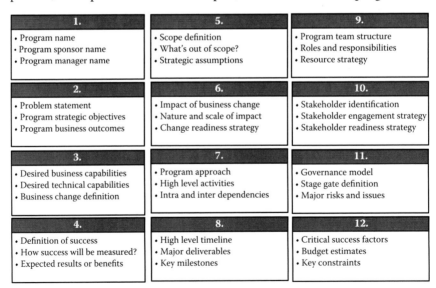

1.	**5.**	**9.**
• Program name • Program sponsor name • Program manager name	• Scope definition • What's out of scope? • Strategic assumptions	• Program team structure • Roles and responsibilities • Resource strategy
2.	**6.**	**10.**
• Problem statement • Program strategic objectives • Program business outcomes	• Impact of business change • Nature and scale of impact • Change readiness strategy	• Stakeholder identification • Stakeholder engagement strategy • Stakeholder readiness strategy
3.	**7.**	**11.**
• Desired business capabilities • Desired technical capabilities • Business change definition	• Program approach • High level activities • Intra and inter dependencies	• Governance model • Stage gate definition • Major risks and issues
4.	**8.**	**12.**
• Definition of success • How success will be measured? • Expected results or benefits	• High level timeline • Major deliverables • Key milestones	• Critical success factors • Budget estimates • Key constraints

FIGURE 2.6
Program charter: defines the problem and sets expectations.

deliver. The next four building blocks of the program charter expound the program scope, the impact of the program on the business, the program approach to attain the desired future state, and the high-level timeline for the program, including key deliverables and milestones. The building blocks 9 and 10 detail the program organization, roles, responsibilities, authority levels, resource strategy, program stakeholders, and stakeholder engagement strategy to manage stakeholder expectations throughout the program. The last two blocks of the program charter will elaborate the program governance model, major known risks and issues, what is needed to make the program successful, constraints that limit the program, and the program budget.

Case Study: Software Licensing Transformation Program

CONTEXT

A Fortune 50 high-tech software company with global operations had designed and implemented a volume licensing program for its top tier enterprise customers. The roll-out of a new family of software products by the company and the procurement of the same by the enterprise customers led to increasing complexity of the licensing program.

BUSINESS PROBLEM OR OPPORTUNITY

The company's top-tier customers were large enterprises with high volume purchases of multiple software products. The customers were in need of better processes, educational collateral, automated tools, and reports pertaining to the volume licensing program, all of which would enable them to maximize the benefit from their purchases. The customers were challenged in administering the licensing program, and the learning curve to ensure compliance with the company's licensing program was steep. The customer inquiries and support needs pertaining to the software company's volume licensing were not being effectively handled by the company's volume licensing functional organization and the company's resellers. The lack of automated tools from the software company that customers could use to track software purchases of all their products impacted the customer experience and interjected inefficiencies, as the customers had to track licensing manually for certain products.

SOLUTION

As the large top-tier enterprise customers accounted for high revenue for the software company and had unique needs, the software company sponsored and embarked on a large program to transform the volume licensing function. The program team led a Voice-of-Customer study with the company's top-10 enterprise customers on their volume licensing experiences to identify and prioritize the customer needs and pain points. The program manager conducted an environment scanning exercise to assess the company's software volume licensing practices against the licensing trends in the software industry.

The volume licensing transformation program charter elaborated the strategic business objectives, defined the problem to overcome, and articulated the definition of success. The findings and recommendations from the VoC study and the environment scanning exercise was used by the program management team to develop and secure sign-off on the program charter from the three sets of stakeholders—customers, resellers, and internal functions. The business outcomes and benefits to be realized by the transformation program were defined from a customer, internal operations, employee, and financial perspective. An integrated end-to-end program plan was prepared, which included projects to tackle "quick wins."

BUSINESS OUTCOMES AND BENEFITS

The implementation of the identified and socialized "quick wins" aided in improving the customer experience of the top-tier customers. An example of a quick win was the expansion of the monthly forum for customer education to introduce the newly developed volume licensing educational modules that addressed specific customer support needs. The volume licensing transformation program team reverse engineered an existing tool to incorporate the licensing quote functionality for all of the missing software products. The modified, enhanced tool assisted the customers in making the best software purchasing and licensing program decisions that met their functionality needs and budgetary constraints.

The transformation team engaged with the internal operations team in streamlining processes to track software usage against the software purchases. The deployment of the streamlined processes to support the volume licensing program not only enabled the large enterprise customers of the software company to easily track compliance, but also benefited the company by minimizing the loss of licensing non-compliance. The software company's early, focused effort in describing and validating the problem and pain points positioned it to deliver the desired business outcomes and realize the expected benefits.

SUMMARY

The program management team plays a critical role during the early stages of the program by ensuring that there is due diligence and governance around business problem definition and alignment among the stakeholders on the problem that the business transformation program will solve. The environment scanning and the voice-of-customer techniques touched upon in this chapter assist in identification, definition, and validation of the business problem. Effective initial and ongoing communications are needed to reinforce the business problem the transformation program has been designed to solve. Clarity around the problem statement, strategic business objectives, and desired business outcomes is a "must have." A common understanding and agreement of the problem statement among the stakeholders creates an invaluable platform that the program management team can further build upon as they successfully drive the program forward.

REFERENCES

Denove, Chris, and James D. Power IV. 2007. *Satisfaction: How every great company listens to the voice of the customer.* New York: Penguin Group, Portfolio Trade.

Levin, Ginger, and Allen R. Green. 2013. *Implementing program management: Templates and forms aligned with the standard for program management and other best practices.* 3rd ed. Boca Raton, FL: CRC Press.

Meyer, Christopher. 1994. How the right measures help teams excel. *Harvard Business Review*, (May-June): 95–103.

3

Articulate the Program Vision and Objectives

The previous chapter discussed the need to obtain an accurate definition of the business problem statement and described two techniques to accomplish this. The upfront clarity and stakeholder agreement on the problem the business transformation program is designed to solve will pave the path for program success. An additional critical success factor is validating and communicating that the problem being solved by the transformation program will position the organization to realize the program vision. Throughout the program life cycle, it is important to showcase to the stakeholders how the program mission, vision, and strategy are constantly aligned to the enterprise mission, vision, and strategy.

The initial and ongoing alignment of the transformation program to the enterprise strategy will result in accomplishment of the transformation program objectives, which is correlated to the enterprise strategic objectives. The program manager should make sure that the program vision and program objectives have been crafted and effectively shared. The upfront and ongoing periodic communication of the desired business outcomes that the program is expected to deliver will enable the program team to stay laser-focused on those work components of the transformation program that will drive the achievement of the business objectives.

The following topics are described in this chapter with the help of supportive illustrations, including a real world case study:

- Formulation of business strategy
- Business transformation drivers
- Implementation of business strategy
- Comparing program management life cycle to the PMI (2013a) *Standard for Program Management*

- Strategic alignment technique: Overview, objective, and approach
- Business performance calibration technique: Overview, objective, approach, and critical success factors
- Strategic imperatives architect programs
- Benefits realization strategy
- Management of benefits realization life cycle
- Case study: Transformation program to redesign process and technology

FORMULATION OF BUSINESS STRATEGY

Business strategy defines how the enterprise/business unit/business area will succeed in terms of objectives and goals. The enterprise vision is the desired future state for the business. The enterprise mission is the highest level statement of purpose for an enterprise, and it provides the identity. The mission statement communicates "who we are, what we do, and where are we heading." The vision and mission drive the formulation of business strategy, which has a longer time horizon. To achieve the objectives laid out by the business strategy, organizations create a tactical strategy to facilitate realization of the near-term objectives and corresponding target business outcomes.

The available, proven technologies and emerging new technologies are constantly forcing enterprises to assess how they can best tap into technologies so that they can become an innovator and market leader. The innovation-driven transformation enables organizations to continuously perform better as they pursue attainment of the articulated mission. After formulating the business strategy, enterprises devise technology and operational strategies. Business objectives can be decomposed into specific operational and technology objectives. The formulated operational strategy focuses on achievement of the operational objectives, and the technology strategy enables realization of the technology objectives.

At the highest enterprise level, organizations need to define the capabilities they need to realize the vision, and this definition is done through the development of an enterprise architecture. The organizational capabilities to pursue the mission can be broken into desired business and technical capabilities at the next level. Business architecture (which is a component of the enterprise architecture) describes the desired business capabilities as a series of cohesive and aligned building blocks, and the technical architecture (which is the other component of the enterprise architecture) does the same for the desired technical capabilities.

BUSINESS TRANSFORMATION DRIVERS

Organizations need to constantly innovate in the current market environment of rapidly changing customer needs, higher competition, increasing globalization, and improved capabilities offered by new technologies. The criticality of improving the business performance level—offering differentiated products and services that address the unmet customer needs and faster response times—is high for maintaining organizational success. The following list provides examples of external and internal drivers that are making organizations embark on strategic initiatives that will transform the current state of business upon their successful execution:

External drivers
- Increased customer satisfaction
- Differentiated competitive positioning
- Improved relationship with business partners
- Decreased time to market

Internal drivers
- Standardization of processes, methods, and systems
- Risk mitigation
- Decreased costs
- Resource optimization

IMPLEMENTATION OF BUSINESS STRATEGY

The initial and ongoing alignment of the operational and technology strategies and the supporting corresponding architectures is critical, and this alignment outcome is accomplished through the business planning and technology planning processes that are in place within these organizations. The defined business architecture can be realized though the implementation of the business road map, which is a high-level plan of the operational work to be completed to achieve the organization's goals. Similarly, the technical road map draws out the technology work that needs to be accomplished. The time horizon will dictate the level of detail in the road map. The critical success factor is to ensure that the business and technical road maps are in sync. The design and implementation of the business and technical architectures will position the organization to realize the desired business and technical capabilities. The business road map, technical road map, program architecture, and program management life cycle are the backbone for the implementation of business strategy.

Figure 3.1 highlights a framework that an enterprise can use to mobilize around its operational and technical strategies and align the strategy through the glue of program architecture. The portfolio management work undertaken at the enterprise level defines the strategic initiatives and establishes the transformation program(s) to manage those initiatives. The program vision is realized based on the successful execution of the transformation program. Chapter 7 provides a comprehensive elaboration on program architecture and how it drives the program management life cycle.

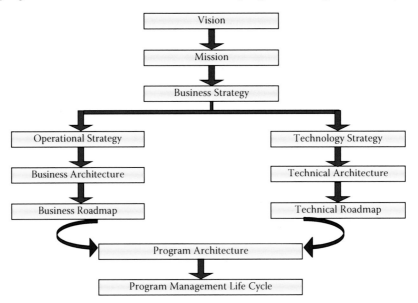

FIGURE 3.1
Top down approach drives strategic alignment.

TABLE 3.1

Comparison of Program Management Life Cycle and PMI Standard

Book	PMI Standard
Four Phases of Program Management Life Cycle	**Three Phases of Program Life Cycle**
1. Set the stage	1. Program definition
1.1 Formulate a program strategy	1.1 Program formulation
1.2 Develop a program road map	1.2 Program preparation
2. Decide what to do	2. Program benefits delivery
2.1 Define the program charter	2.1 Component planning and authorization
2.2 Create the program plan	2.2 Component oversight and integration
	2.3 Component transition and closure
3. Make it happen	3. Program closure
3.1 Execute the program	3.1 Program transition
3.2 Monitor the program delivery	3.2 Program closeout
4. Make it stick	
4.1 Transition to operations and close the program	
4.2 Sustain outcome delivery	

A transformation program is executed and managed by the four-phased program management life cycle, which was introduced in Chapter 1. In the next section, Table 3.1 maps the terminology used in this book to the terminology used by the Project Management Institute (PMI 2013a).

COMPARING THE PROGRAM MANAGEMENT LIFE CYCLE TO THE PMI *STANDARD FOR PROGRAM MANAGEMENT*

The program management life cycle detailed in this book is not identical to the "program life cycle" in the PMI's third edition of *The Standard for Program Management*. In the current book, each of the four phases comprises two processes, and there are a total of eight processes. In the PMI standard, the three phases comprise subphases, and there are a total of seven subphases. A high-level mapping is provided in Table 3.1 to draw out the distinctions and some commonalities in the terminology. Please note that matching numbers in Table 3.1 do not convey that the two are identical, i.e., "Set the stage" is not the same as "Program definition."

STRATEGIC ALIGNMENT TECHNIQUE

Overview

Strategic alignment is the highest level thinking and analysis used to align the transformation program strategy to a business strategy. As the business strategy drives the operational and technology strategies, the transformation program strategy has to be aligned with these strategies. The program management team can facilitate the macroanalysis exercise that validates and ensures strategic alignment by involving and engaging the senior leaders. In addition, such an exercise establishes the linkage between strategy and implementation of that strategy through the program management life cycle. Chapter 7 highlights how the program management life cycle and the underlying program management processes drive the implementation of the strategy. Chapter 7 also covers the development and execution of the comprehensive transformation program plan to achieve the strategic business objectives and realize the targeted outcomes.

Objective

Macroanalysis provides the "big picture" view of the transformation program. The alignment of the transformation program to the strategic direction of the enterprise through the course of the program life cycle can be confirmed by completing a strategic alignment exercise (Figure 3.2). As part of the strategic alignment work, the program management team develops the program mission, program vision, program stakeholder values, and program objectives. These need to be communicated to all the stakeholders.

Approach

The alignment of the business transformation program to the business strategy is accomplished by performing the following steps:

- Create the program vision
- Develop the program mission
- Determine the program stakeholder values, including values that the core program team will use
- Develop the program objectives and goals

Program Vision	"To be the premier provider of preventive and non-preventive health care solutions and services in each of our markets"		
Program Mission	"To help our members maintain high quality health"		
Program Objectives and Goals	We will accomplish our mission by working with our members to: 1. Fulfill their expressed and implied health care needs by integrating and leveraging the spectrum of health care solutions and services 2. Deliver the ideal experience as appropriate for each member 3. Lower cost of health care to members without curtailing benefits		
Transformation Program	1. Development of integrated health care model to better fulfill health care needs by integrating across payer, provider, and physician	2. Enhancement of member experience by voice of the customer, member segmentation, consistent policies, and satisfaction monitoring	3. Creation of an operationally excellent environment by streamlining business processes, information systems, and tools

FIGURE 3.2
Directional alignment framework (sample for a health care company).

Step 1. Create Program Vision

The program vision is the business end state to be achieved by a transformation program. The program vision is the final output of the iterative visioning exercises among the senior executives, portfolio review board or other executive oversight group, program steering committee, and program sponsor. The transformation program team and stakeholders need to share a common understanding of the defined business problem, what they are trying to accomplish, and the desired end state. This unified understanding is best accomplished through the program vision statement. The program manager assists the program sponsor in creating the program vision collateral, which includes the vision, mission, values, and objectives. There has to be an alignment on the program vision between the program team and the executives accountable for realizing that vision. Chapter 7 further elaborates on the definition of the program vision and the program future state.

Step 2. Develop Program Mission

The program mission is the highest level statement of purpose for the program, and it aligns with the enterprise mission. The program mission statement should clearly communicate to all transformation program stakeholders "who we are, what we do, and where are we heading." The program team needs to understand the program mission, which should

- Enable the program to avoid visionless or rudderless leadership
- Convey a purpose and identity that motivates the program team and stakeholders
- Provide a context for making strategic decisions during the program life cycle

Step 3. Determine Program Stakeholder Values

Transformation program stakeholders are individuals and groups, both internal and external, who have a stake in the success of the program or who are affected by its actions. Examples of program stakeholders are customers, employees, shareholders, and business partners. Program stakeholder values guide the actions and behaviors of everyone involved in the program, and these are documented in the stakeholder value statements. The practice is to develop such statements for each stakeholder group and to make it germane to that group. As these statements become a kind of constitution for the transformation program, they are communicated and reinforced by the program manager. A few examples of program stakeholder values are: open communications, respect, accountability, honesty, and collaboration. A sample program stakeholder value statement related to "open communications" would be, "Program team members should look for and appreciate constructive feedback."

Step 4. Develop Program Objectives and Goals

Program objectives are the documented statements of purpose that support the program mission statement. They are permanent and represent the highest level and long term goals of the program. The short term and medium term goals the program needs to deliver should also be defined by the program management team. Program objectives are more detailed than the program mission, generally addressing the specific qualities and goals that promote and contribute to the achievement of the program mission. They are typically targeted at one or several of the stakeholder groups within the program, e.g., offering world class service to customers. One of the significant aspects of upfront determination of program objectives is that they provide a basis for measuring the business performance and assessing the success of the transformation program. Figure 3.2 provides a sample illustration of vision, mission, and objectives of a transformation program.

With the constantly changing competitive landscape, launching of new products and services, increasing customer expectations, and emerging technology platforms, enterprises have to assess and monitor the external forces and take timely strategic actions. Highly successful enterprises that are dominating markets must do the same if they are to continue to be the leaders. On the other hand, the laggards want to keep improving or capitalize on any missteps by the market leaders to become the leaders. As a result, enterprises are designing strategic initiatives (or transformation programs) whose scale may vary from an enterprise level to line of business function.

In this day and age of scarce resources and the need to get more done with less, the portfolio management function assists in prioritizing the launch of strategic initiatives. The program management discipline drives the end-to-end execution of the launched transformation program or strategic initiative and enables achievement of the agreed upon program objectives. A research paper by PMI (2013b), "The Impact of PMOs on Strategy Implementation," shows that aligning programs and projects with strategic objectives has the greatest potential for adding value to the organization.

BUSINESS PERFORMANCE CALIBRATION TECHNIQUE

Overview

Enterprises or business areas within the enterprise use the Business Performance Calibration (BPC) technique to compare and assess aspects of their business performance against those of other organizations. The BPC is the continual process of searching for and analyzing, understanding, and measuring the best business practices. The implementation of these best practices as part of execution of the transformation program will achieve a superior level of business performance. The transformation program vision, mission, and strategic objectives coupled with the BPC findings aids in developing a program plan that has factored the best practices.

The successful execution of the program plan by the program team results in the attainment of the future transformed end state whose business performance is at par with or better than the best in the industry. The BPC can also provide the basis for the business change the transformation program is driving and garner the support of program stakeholders. In other words, BPC orchestrates the delivery of measurable, tangible benefits as wells as intangible benefits, like keeping the core program team motivated to develop the best outcomes.

BPC is not an exact science, but an art. Each BPC study has its own unique information needs. The time it takes to complete BPC is variable, and in many instances a vast amount of business insight and judgment have to be used. Key first questions to ask are:

- How is the organization performing (not just financially)?
- How to determine if the organization is performing to the best of its abilities?
- How is the competition performing in selected areas?
- What are the primary factors that result in customer satisfaction?
- What processes or activities and their associated key value indicators can help in making sound strategic decisions?

Objective

BPC enables enterprises to become the "best of the best." This technique is used to discover what practices are needed to reach higher performance levels and higher targets. BPC can be used to:

- *Focus on customer requirements*: The external focus ensures that industry best practices will be uncovered and helps the program in exploring ways to improve customer satisfaction.
- *Compare performance externally and internally*: The results of this "best practices" comparison are assessed to determine what processes need innovation or need to be adopted.
- *Check how enterprise strengths and weaknesses stack up against the competition*: This comparison can show where competitive opportunities exist and clarify core competencies of an enterprise, which in turn helps in breaking down the "not invented here" syndrome. In their *Harvard Business Review* article, Tucker, Zivan, and Camp (1987) note that comparisons with competitors may uncover practices for meeting competitor performance, but not reveal practices for beating them.

- *Provide a proactive way to effect change*: Enable the program stake-holders to better understand their outputs and appreciate how other enterprises do things well.

Approach

The four generic phases of BPC are:

- Planning
- Analysis
- Articulation
- Implementation

Phase 1: Planning

The planning step is particularly critical. In this step, the needs and uses for BPC are defined, as without these goals in place, the BPC effort will be ineffective. The objective of this step is to determine the "what" and "who" to calibrate and how the data will be collected. When deciding "what" to calibrate, the high impact business areas should be chosen to ensure the timely and efficient use of resources and to confirm that only relevant aspects of those business areas are calibrated. In deciding "who" to calibrate, consider the industry leaders in a specific area, competitors, customers, and organizations outside the industry. Usually, calibration subjects include functions, processes, and activities:

- With a strong influence on business performance
- Impacting customer perception of the business
- That separate the organization from its competition
- That are essential to the business

Phase 2: Analysis

This BPC phase involves analyzing the information and data gathered in the planning phase. There are three types of business performance gaps:

- *Negative*: This gap indicates that external practices are superior and that the performance baseline should be based on external practices. A major effort is required to determine what must change in order to change internal practices.

- *Positive*: This gap indicates that internal practices are clearly superior and should serve as the performance baseline and receive appropriate recognition.
- *Parity*: This gap shows no significant differences in practices. Further analysis is required to determine the appropriate methods that will lead to superior practices.

The following steps are needed to complete the performance gap analysis:

- Identify the gap and tabulate both descriptive and numerical data
- Assess and describe reasons for the quantitative gap
- Evaluate and determine reasons for the qualitative gap
- Evaluate key factors leading to best practices, i.e., processes, standards, culture, etc.
- Determine the performance baseline

Phase 3: Articulation

The following steps are to be completed for the articulation phase of BPC:

- Communicate the findings
- Obtain executive leadership support on the performance baseline
- Redefine objectives, outcomes, and key value indicators (KVIs) to close the performance gap

Phase 4: Implementation

The final phase of BPC is the conversion of findings and recommendations into a business performance improvement plan. BPC enables the organization to define changes that must be made. The implementation phase thus converts the redefined objectives and outcomes into specific implementation actions and puts into place a periodic calibration mechanism. The following steps are performed in BPC implementation:

- Develop action plans and targets
- Implement specific actions
- Monitor outcomes against performance baseline
- Communicate the realized outcomes

Critical Success Factors

- A clear understanding of "what" is going to be calibrated
- Recognition that BPC findings can be misleading if the comparison of the best practices is not right.
- Caution has to be exercised while comparing a business process with an identical business process in an altogether different industry.
- A willingness to change and adapt based on BPC findings
- Active commitment by management to the process
- A realization that BPC is a continual process
- Adherence to a structured approach to BPC
- Openness to new ideas and creativity as well as innovation in their application to existing business processes

STRATEGIC IMPERATIVES ARCHITECT PROGRAMS

The calibration exercise for business performance provides invaluable information to enterprises on the business environment in which they operate. Key insights on market trends, competitor strategies, operational best practices, etc., become apparent. Because of the deliverables and outcomes resulting from the strategic alignment related work, the strategic direction is known and the executive sponsorship is confirmed. At this point, business, operational, and technology strategies will have been formulated, and these strategies will have taken into consideration the factors that are driving the strategic direction of the business. These strategies now need to be taken forward and implemented.

Strategic imperatives can be derived by the program management team based on environment scanning, business drivers, organizational performance levels, and organizational capabilities. Strategic imperatives identify the innovation, improvement, and business change opportunities for the enterprise. Program management can own the planning and execution of these imperatives. A sample mapping of business drivers to strategic imperatives is shown in Figure 3.3.

The strategic imperatives are typically overarching, complex, and high level, and each can comprise multiple large building blocks. The portfolio

FIGURE 3.3
Formulate program strategy by aligning to business drivers.

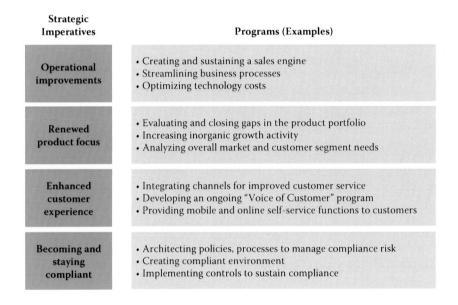

FIGURE 3.4
Design program by mapping to strategic imperatives.

management function can analyze the strategic imperatives, identify these large building blocks, and logically group these building blocks based on commonality, interdependency, and correlation. These logically grouped building blocks are programs. Figure 3.4 provides a sample mapping of strategic imperatives to programs.

BENEFITS REALIZATION STRATEGY

The program strategy formulated at the enterprise (or business division or function) details a high-level approach to be taken by the program to deliver the targeted future state. The program strategy provides the vision, which is critical, as the lack of a clearly defined and agreed future state on the culmination of the program results in significant challenges and barriers. The program vision is realized through successful implementation of the program strategy. The program strategy per se does not achieve the business objectives and benefits expected of the program. The benefits realization definition process identifies and defines the tangible and intangible benefits the program is expected to deliver and sustain.

The benefits realization strategy spells out the approach the transformation program will take to achieve the business objectives, deliver the expected outcomes, and realize the business benefits. As part of the benefits realization strategy work, it is a best practice to delineate the business outcomes and expected business benefits into short-term and long-term results based on the time horizon needed to realize them. In the early stages of launch of the transformation program, the benefits realization strategy is articulated. The execution of a benefits realization strategy results in the attainment of program business objectives, delivery of expected outcomes, and realization of program benefits. The program manager validates the business benefit expectations of program stakeholders and the program sponsor and secures their sign-off on the benefits realization strategy.

BENEFITS REALIZATION LIFE CYCLE

The focus of the program management discipline is to drive the business transformation forward by planning, executing, monitoring, and transitioning the launched program. Chapter 1 described the program management life cycle. Figure 3.5 showcases how the benefits realization life cycle stages are embedded within the program management life cycle. Program management drives the benefits realization life cycle, beginning with a benefits definition stage and ending with a benefits sustainment stage. The four stages of benefits realization life cycle are

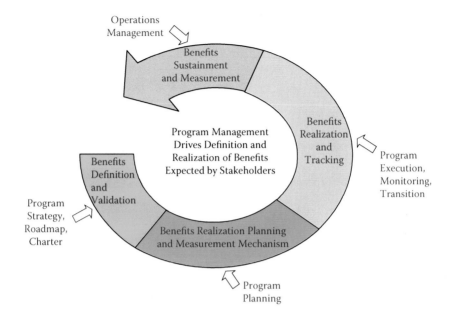

FIGURE 3.5
Benefits realization life cycle.

- Benefits definition and validation
- Benefits realization planning and measurement mechanism
- Benefits realization tracking
- Benefits sustainment and measurement

The successful execution of the program management life cycle by the program team and transition of the program to the designated operational units results in the achievement and sustainment of the targeted business outcomes and business benefits.

Case Study: Transformation Program to Redesign Process and Technology

CONTEXT

A leading midsize metals manufacturing company and producer of aluminum products wanted to capitalize on the tremendous market opportunity to be a more dominant player in a niche market. The lack of clarity around the company's strategic direction and the inability to scale the internal infrastructure posed a significant challenge in realizing the vision.

BUSINESS PROBLEM OR OPPORTUNITY

The manufacturing company was structured as autonomous lines of business for each of the product lines. Some of the product lines needed a discrete manufacturing model, while the others needed a repetitive manufacturing model. The company's senior management had not clearly articulated the organization's vision and the strategic direction of the company. Although the company's unique capability to accommodate the different manufacturing models was a true differentiator, it lacked the blueprint to scale these models and the supporting processes and systems. The operational processes within each line of business were highly tailored, which worked well, but standardization across the lines of business was a monumental challenge. The company lacked the integrated information systems that would be needed to scale operations to meet the demand.

SOLUTION

The business performance of the metals manufacturing company had been consistent over the years, and it had the strong financials to support the investments needed to grow. The senior executive team proposed and secured the board of directors' approval to redesign the business. One of the first steps taken by the skeleton redesign program team that was put in place was to lead a series of workshops to land on a unified vision. The team applied the strategic alignment technique in formulating the business strategy by engaging with the senior executive team and the leaders within each line of business.

The strategic direction of investing in and supporting both the manufacturing models was agreed upon and communicated to all the stakeholders of this business transformation initiative. The size of the core program team was expanded, and the team completed a business performance calibration exercise to get a deeper understanding of the operational processes and information systems in vogue at other similar companies within and outside of the metals manufacturing sector. The program manager developed and socialized the benefits realization strategy prior to launching the next phase of the transformation program. The next phase entailed redesign and standardization of business processes within the lines of business and deployment of new integrated systems that support multiple lines of business.

Business Outcomes and Benefits

The initiated business redesign program to "grow and get better" had a common understanding of the vision across the key executives and stand-alone lines of business. The business performance calibration exercise generated new ideas on how other comparable organizations had gone about in scaling their internal operations and systems to accommodate the strategic objective of business growth. The program team secured stakeholder buy-in and executive sponsor agreement to implement new enterprise-strength information systems. The standardized operational processes improved operational efficiencies, and the redesigned scheduling methods for the shop floor increased the capacity of manufacturing. The articulation of the company's redesign program vision and strategic objectives to be accomplished set the stage for success by driving alignment across the lines of business and numerous functional areas.

SUMMARY

Programs that are designed to transform a business are complex and high risk, as the magnitude of change being driven is high. As the program management team drives forward the planning and execution of programs, all of the program stakeholders need to be cognizant of the program vision, program mission, program stakeholder values, and program objectives and goals. In addition to awareness of the program vision, for true business change to happen at all levels, there has to be a buy-in to the vision. The strategic risk of program execution without comprehending program vision is similar to the risk of program execution without adequate program planning. Periodic reinforcement of the program vision and of how achievement of the program objectives will benefit everyone is one of the most important critical success factors for a transformation program.

REFERENCES

PMI. 2013a. *The standard for program management.* 3rd ed. Newtown Square, PA: Project Management Institute.

PMI. 2013b. *The impact of PMOs on strategy implementation.* Pulse of the profession in-depth report. Newtown Square, PA: Project Management Institute. http://www.pmi.org/~/media/PDF/Publications/PMI-Pulse-Impact-of-PMOs-on-Strategy-Implementation.ashx

Tucker, Francis, Seymour Zivan, and Robert Camp. 1987. How to measure yourself against the best. *Harvard Business Review,* (January-February): 137–46.

4

Secure Cross Functional Executive Sponsorship

Sponsorship is the single most important factor in ensuring successful transformation of a business through program management. Numerous studies on transformation programs have shown that one of the major obstacles for transformational change is the lack of executive commitment to the program. For a complex program to be successful in realizing the program vision, achieving business objectives, and delivering the targeted business outcomes, the need for an executive champion through the course of the program is critical. Sponsorship is this executive commitment and championship, and typically this role is titled *executive sponsor.* The program executive sponsor authorizes, legitimizes, and demonstrates ownership of the program. Since the program management team confronts sticky cross divisional and cross functional issues, risk, changes, constraints, and obstacles, the program manager ensures that the program gets the highest level sponsorship from the executive sponsor throughout the program management life cycle. Program sponsorship is needed at multiple levels and in varying degrees at different stages of the program to change the business. The program manager ensures that the sponsorship needs are being met from program start to end.

The following topics are described in this chapter with the help of supportive illustrations, including a real world case study:

- Sponsorship of business transformation program
- Multilevel program sponsorship model
- Program value justification technique: Overview, objective, approach, and helpful hints
- Upward management of program sponsors

- Sponsorship of program outcome delivery and benefits realization
- Case study: Program to transform procurement function via outsourcing

SPONSORSHIP OF BUSINESS TRANSFORMATION PROGRAM

Most transformation programs are complex, as they drive significant business change across organizational boundaries and impact a large number of stakeholders. The business change definition and impact analysis work of the program team coupled with the stakeholder analysis provides the key inputs to develop the sponsorship model. The example in Figure 4.1 highlights the nature and layers of sponsorship needed on a typical transformation program.

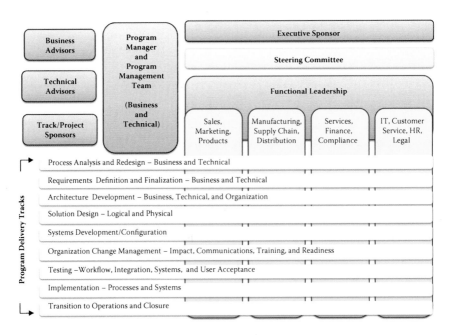

FIGURE 4.1

Sponsorship model for a typical transformation program.

The program sponsorship model is designed to have the vertical (or functional) sponsorship as well as the horizontal (or track) sponsorship. This program is impacting stakeholders in numerous business functions (e.g., sales, supply chain, and finance), and functional level sponsorship is necessary to either comply with or commit to the changes within those functions. The cross functional changes that this program is driving can be implemented with executive sponsorship, as that organization level possesses sufficient power and influence to do the same. The business change the program is driving is supported by the program steering committee comprising influential stakeholders from the impacted business areas.

The illustrated program has numerous horizontal delivery tracks to facilitate execution of this program. Each of these tracks is led by project manager(s), and each track is ultimately owned by a track sponsor. The program manager and track sponsor collaborate in ensuring the highest level support to the program delivery tracks. So, if the team tasked with tracking the business and technical requirements is encountering any obstacles, the program manager can support that team by eliminating the barriers.

Typically, the program manager plays the sponsor role for the project(s) in the delivery tracks. Based on the organization's structure, funding mechanisms, and responsibility boundaries, it is possible for a nonprogram manager to play the project sponsor role for a subset of the projects constituting the program. However, in such a scenario, the sponsoring individual partners with the program manager are directing and supporting that project subset. The transformation program manager ensures that the ultimate goal—getting the right stakeholders to engage in the right program activities at the right times—is achieved on the program.

The delineation of roles and responsibilities within the core program team is essential. Figure 4.2 highlights the following six roles in the context of the sample transformation program presented here:

- Executive sponsor
- Steering committee
- Program manager
- Track sponsor
- Project manager
- Core team

Role	High-Level Responsibility Description
Executive Sponsor	Articulate mission and vision for the program. Provide leadership. Make strategic decisions on program direction. Ensure alignment to strategy. Approve large funding requests.
Steering Committee	Share strategic direction and objectives. Determine priorities. Make program decisions. Provide leadership and resources. Address escalations from Program Manager. Ensure synergy with other cross programs.
Program Manager	Drive program plan, review of deliverables, and change control. Track dependencies, risks, and issues. Resolve issues. Provide update to Steering Committee and Executive Sponsor on program progress.
Track Sponsor	Articulate vision for the track. Partner and collaborate with the Program Manager. Provide track level subject matter expertise to the team. Provide leadership and reinforce commitment to the track.
Project Manager	Provide update to Program Management on track progress. Develop and drive project plan. Review project deliverables. Assess project level risks, issues, and change control. Resolve and escalate.
Core Team Member	Execute in line with the project and program plan. Collaborate with others on dependent tasks. Provide subject matter expertise. Relay work progression status and areas needing attention.

FIGURE 4.2
Roles and responsibilities of the transformation core team.

MULTILEVEL PROGRAM SPONSORSHIP MODEL

Multilevel sponsorship is about having sponsors at different levels of the program and business organization hierarchy. There is the main, primary sponsor at the highest level of hierarchy for the impacted organization, and this sponsor is supported by additional secondary and tertiary sponsors at the lower organizational levels of the impacted organization. The secondary and tertiary sponsors reinforce and model the actions and behavior of the primary sponsor, and all these three levels of sponsorship have to be aligned.

A best practice is for each level of the program organizational chart to have sponsorship, and hence the key leader at each level needs to play a sponsor role by expressing and reinforcing his or her commitment to the program. For example, program managers need to provide program level sponsorship to the project managers, and the project managers need to

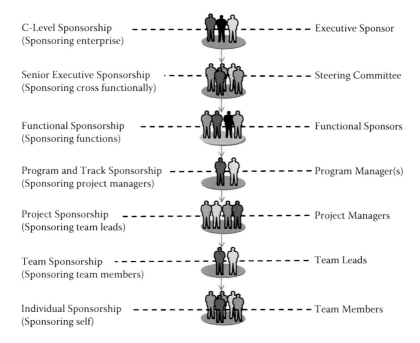

FIGURE 4.3

Multilevel sponsorship needed for success of transformation program.

provide project level sponsoring to the team leaders. As sponsorship is also about owning and embracing, an individual's commitment to the program is his or her sponsorship. The sponsorship outcomes that the three sponsor types can deliver will vary, as they are at different hierarchical and authority levels with varying spheres of influence. Figure 4.3 showcases the multilevel sponsorship model from a different angle.

PROGRAM VALUE JUSTIFICATION TECHNIQUE

Overview

Justifying program value is one of the techniques available in the program management tool kit that enables development and maintenance of the business case to secure and sustain the sponsorship for the transformation program. The executive sponsor proposes the transformation program to the organization's senior executive team (e.g., board of directors, C level executives, or portfolio review board) along with the business case for justification. Program value justification (PVJ) is a set of compelling arguments detailing the value of implementing business change driven by the program.

PVJ ties the value of the transformation initiative to its ability to achieve its enduring strategic business objectives. It usually has both a monetary and nonmonetary component, and includes an analysis of both costs and benefits. PVJ also includes the analysis of risk and return in order to provide a complete picture of the value of the program to the stakeholders and the enterprise. PVJ analysis needs to be flexible, as the modeling of different scenarios is important to develop analysis that shows both the expected return and the level of risk. PVJ may take the form of a spreadsheet, a series of bullet points, pages of scenario analysis, or all of these, depending on program complexity, size, and sponsorship needs.

Objective

The purpose of PVJ is to make the transformation program selection and program continuation processes as effective as possible so that the correct business decisions are made quickly and with a high level of consensus. During the program selection stage, the primary audience for PVJ results is the senior executive team of the organization that makes the program selection and funding decisions. For the continued funding of a transformation program that is in flight, the primary audience for PVJ results is the designated executive sponsor of the program. PVJ, regardless of format, needs to provide the targeted audience with the information needed to make informed decisions.

At a minimum, the PVJ has to provide financial costs and expected business results information to enable the executives to make program funding decisions. In many instances, however, executives also need softer measures of change (e.g., positive publicity from the roll-out of an innovative service) to complete the real picture of the value of business outcome the program will realize. Two additional examples of intangible benefits to be taken into consideration in making program funding decisions would be: higher customer satisfaction from the launch of a customer data protection initiative and improved customer experience from usage of mobile technologies.

For stakeholders not directly involved in the business transformation program, PVJ is often an excellent vehicle for demonstrating the value of the changes they will be experiencing. If employees will see their jobs change, it is important for them to understand why their jobs will be changing. If they can see the change in terms of expected improved customer service with a dollar value attached to it, they are more likely to buy into the changes than if they are told, "The executive committee made a decision; deal with it."

Approach

The important exercise of development of the PVJ deliverable involves seven steps:

- Assemble key input deliverables
- Describe future work
- Determine cost components of the program driving transformational business change
- Determine tangible benefit components of the program driving transformational business change
- Determine intangible benefit components of the program driving transformational business change
- Measure risk and return
- Formalize PVJ

Step 1: Assemble Key Input Deliverables

The purpose of PVJ is to demonstrate the value of designing and implementing the innovations developed within the program. The key inputs are the results of strategic alignment, impact analysis, performance improvement measurement, and stakeholder analysis. Impact analysis should contain the financial costs and benefits of innovations. Performance improvement measurement measures and evaluates the improvement in business performance as a result of delivery of the planned business benefits by the program. Communication plans are essential, as the needs of each set of stakeholders should be addressed by the PVJ.

Step 2: Describe Future Work

The initially developed high-level cost estimate has to be validated and refined as the program plan gets fully developed. The program plan and project plans that have already been created may or may not contain the information necessary to calculate the costs of the program. The program plan information necessary to calculate program costs includes the duration of each program plan task, the resources necessary for each task, and the fully loaded costs associated with the resources. In addition to these costs, the program plan should include program implementation and training activities that will help determine the expected rate of change from the old to the new. The extent and rate of business change has a significant impact on the business outcomes resulting from program implementation. For instance, if an implementation occurs in January but training continues through March, and the old information systems are retired in June, the expected performance levels will not be reached until at least June. These performance-related timing factors have to be taken into consideration.

Step 3: Determine Cost Components of the Program Driving Transformational Business Change

Determining the cost components of change is perhaps the most important activity of PVJ and usually consumes the most time and resources. Cost components should be determined by taking the following substeps:

- *Compare performance levels*: The successful execution of the program plan by the program team results in superior business performance in the future state. The business performance calibration technique, covered in Chapter 3, can be leveraged to determine the current and future state performance. The comparison of performance levels can shed light on the impact of the performance gap between the current and expected performance levels.
- *Assigning monetary value to performance differences*: Once these performance differences are determined, translate them into cash flows and capture them, the assumptions, and other pertinent information in a spreadsheet. The spreadsheet will be the basis for the complete financial PVJ analysis, which should result in something similar to an income statement. When completed, it should show income to the business area at the top, costs below, and net income (or the value of the program to the business area or enterprise) at the bottom. Generally, the most difficult part of this analysis is determining the monetary value for improvements in customer service, quality, and organizational and people oriented changes. One method for the quantification of intangible benefits is market research. For example,

based on the market research findings, a point increase in customer satisfaction ratings due to superior customer service could be translated into a certain monetary value.

- *Assigning program costs*: Along with the current and future business area costs, program specific costs must be included to show the real value and return of any initiative. On the same spreadsheet, program costs should be estimated over time. Opportunity costs are often overlooked when assessing the value of a program, and they should not be. For example, the costs incurred by a program to prototype a couple of strategic options would need to be factored into the program costs even if the conclusion from prototyping turns out to be that none of the options is feasible.

Step 4: Determine Tangible Benefit Components of the Program Driving Transformational Business Change

This step closely parallels Step 3. Revenues should be calculated along with costs to determine the total monetary value of the program. Determining the revenue components involves assessing changes in factors that affect revenue over time and then converting those factors into cash flows. It is easy to assume that the business outcomes of the program will be revenue-neutral when, in fact, they will not be. For instance, if innovative customer service processes are realized when the program is completed, their impact on future sales revenue needs to be considered. While revenue increases (or decreases) that are due to operational changes are often difficult to quantify and must be well documented, they are just as important as costs to complete the PVJ. One method for the quantification is the usage of industrial engineering techniques to determine the higher utilization percent resulting from an improved operational process in the transformed future state. The higher utilization results in extra output, and the sales of that extra output generate add-on revenue.

Step 5: Determine Intangible Benefit Components of Program Driving Transformational Business Change

An attempt should be made to translate all business functional changes into monetary terms. Intangibles can be made tangible, and some examples of how this can be done were illustrated in the previous section. Changes to some areas, such as product quality or customer service, may involve making tenuous assumptions in order to convert them to cash flows. For each program business objective, show how the program completion would impact that strategic objective. For example, if the business objective includes raising customer service ratings above 90%, the PVJ should address it by showing how the improvements resulting from successful program execution will raise the customer service ratings above 90%. The business value opportunities presented by improving customer service ratings above 90% may be to increase repeat sales to existing customers and to create a marketing advantage. The threats of not improving customer service ratings above 90% may include losing market share and increasing marketing expenses to counteract negative customer feelings.

Step 6: Measure Risk and Return

So far, the PVJ has focused on the costs of the transformation program and the expected business benefits based on the best estimations of the program management team. In order to present the most accurate view of the value of business outcomes delivered upon program implementation, the PVJ should also include an assessment of the riskiness of the program. Using the income statement prepared previously, create different versions of the income statement for different cases. A specific program may need different categories, but the below set of generic scenarios should be explicitly examined:

- Expected results
- Worst case results
- Best case results
- Do nothing results

Step 7: Formalize PVJ

The PVJ should generally be packaged and presented in three layers as follows:

- A one-page PVJ at the highest level suffices for executives who need only the critical facts.
- A multipage analysis that describes important points behind the PVJ—describing the value of achieving key measurements, making clear key assumptions, and discussing different scenarios and their impact on the PVJ—is an essential report. Executives interested in

the reasoning behind the one-page PVJ summary will find answers in this analysis.

- An appendix or separate report that includes all of the PVJ information, including spreadsheets for all scenarios, is also essential. Program financial analysts or others interested in details will find all the information they need in such a report.

Helpful Hints

Focus the PVJ on the types of justification necessary to communicate with the target audience. It is not always necessary to build a PVJ to the level described above. Sometimes all that is necessary is five bullet points of how the program will support key strategies and that the costs are manageable. For some audiences, dozens of spreadsheets may be in order. For other audiences, pages of analysis of nonfinancial impacts may form the bulk of the PVJ. There is no easy solution for the final form of PVJ analysis, and the program management team should have a vision of the final PVJ deliverable before starting the PVJ process.

Always try to compare apples and apples. Estimates of future costs and revenues are often tied to product or process volumes. In comparing future costs and revenues to historic or current costs and revenues, it is tempting to use optimistic projections based on product or process volumes that may not be realistic. The reliability and accuracy of PVJ is tied directly to the comparability of current data and future assumptions relating to cost and revenue drivers. Ensure that the assumptions driving projected costs and revenues and learning and transition curves are as realistic as possible. Assuming the transformed business state will be operational on Day One is a common mistake that results in negative perceptions of a generally successful business transformation program. The time taken for a business to reach full efficiency upon successful culmination of the business transformation program is a variant, so the expectation of stakeholders has to be set and managed accordingly.

UPWARD MANAGEMENT OF PROGRAM SPONSORS

The success of transformation programs hinges greatly on continued sponsorship by the executive sponsor, so the program management team has to ensure that the information requirements and other known needs of the executive sponsor are effectively addressed. The ongoing visibility of solid program sponsorship will keep the program team focused on executing the program and not worrying about unexpected strategic shifts and a drop-off of support. The program management processes should proactively manage the expectations of the sponsors so that the following critical actions of the sponsor continue as the program execution progresses:

- Approving the release of program funds
- Staying committed to the agreed program delivery timeline
- Avoiding deviations from the planned business objectives and business outcomes
- Sticking to the agreed scope, quality bar, and permissible risk level for the program
- Fulfilling resource needs after submission of justifications
- Reviewing and signing off on key deliverables
- Making timely decisions on critical change requests that highly impact the program

The program manager has to keep the executive sponsor engaged throughout the program management life cycle. Internal and external forces can impact a transformation program at any time. The ability of the program team to adapt and counter such forces under the direction of the executive sponsor is essential for a business transformation program to succeed. If strategic shifts need to be made, the program manager needs to determine whether to add a new project, cancel a project, or recommend canceling the program.

SPONSORSHIP OF PROGRAM OUTCOME DELIVERY AND BENEFITS REALIZATION

As the program management team monitors the critical path of the program, it is constantly identifying and analyzing the major barriers to the success of the transformation program. These major barriers directly impact the achievement of program objectives, delivery of program outcomes,

			Business Outcome Management Tool				
Id	Program Barrier Description	Impact on Business Outcome	Path to Green (PTG)	PTG Due Date	PTG Owner	PTG Status	Escalate/ Inform
1							
2							
3							
4							
5							
6							

FIGURE 4.4

Sponsorship removes barriers and enables delivery of successful business outcomes.

and realization of business benefits. The program manager navigates these obstacles through the "path to green" matrix highlighted in Figure 4.4.

This matrix should not be confused with the program or project level issues or risks' logs, which are the outputs of the issue and risk management processes at the program and project levels. However, the items in this matrix could be the escalated high criticality issues and high severity risks. The input sources for this matrix can vary and can include the program monitoring dashboard, which evaluates the health of the program on a weekly basis based on the seven health assessment criteria of program timeline, funding, resources, scope, quality, risk, and value. The matrix is a management tool for business outcome delivery. The program manager engages with the executive sponsors to get as needed support to eliminate the barriers to the success of the transformation program. The cross functional executive sponsorship drives the delivery of outcomes and realization of benefits.

After analyzing the barriers with the help of the core program team, including the subject matter experts, the program management team develops the "path to green," which is the approach to eliminate the major barriers and reposition the program to deliver the desired business outcomes and realize the expected benefits. Only the major program barriers are listed in this matrix, so all of them have to be expeditiously overcome. To do so, involvement from the sponsorship team is useful to overcome the potential negative impact these barriers will have on the program. After securing the agreement of the sponsor on the strategy, the program team creates the path to green plan, which describes the specific actions needed to eliminate the major barrier.

The sponsor can play a key role in influencing the benefits realization by:

- Being visible and involved
- Creating an empowered work environment
- Reinforcing commitment to the program
- Advising on the strategy
- Releasing communications to the program team stressing the level of urgency
- Tapping into other senior executives who can assist
- Offering any add on support the program team may need

The PMI (2013) white paper on the role of the sponsor concludes that organizations are more effective when they recognize program and project sponsorship as a core competency. The paper further states that the sponsor has to ensure that the organization's leaders are involved and provide visible support to the program throughout the entire life cycle.

Case Study: Program to Transform Procurement Function via Outsourcing

CONTEXT

A well recognized Fortune 50 high tech hardware conglomerate with a global presence wanted to assess how to innovate the way it was procuring goods and services. An organizational restructuring effort coupled with sweeping technology changes led this giant high tech company to pursue an alternative procurement model.

BUSINESS PROBLEM OR OPPORTUNITY

This high tech conglomerate was in the midst of a strategic initiative to deploy a worldwide eProcurement solution under a business service provider model. Under this model, procurement of certain goods would be done by the service provider using the company's business to business marketplace. The solution deployment began to experience problems when the business service provider started customizing the solution. The company's core program team working on this initiative realized that the service provider lacked the capability to customize their eProcurement solution. The business service provider had not dealt with such a large and complex deployment before, lacked the needed eProcurement solution experts, and did not have a dedicated team on this deployment. The high tech company's program management team for this initiative was challenged in legitimizing the projections for reduced technology spending in supporting eProcurement functions and higher savings from the predicted lower procurement cost for goods and services.

SOLUTION

The program management team decided to take a step back and engaged with the company's procurement function leadership as well as the executive team of the business service provider tasked with developing the eProcurement solution for the company. As part of exploring options to turn around this strategic initiative, the team assessed the program sponsorship model to ensure that the right sponsorship structure was in place and that the sponsor needs were being captured and documented. Given the impact of this initiative to many functions outside of the company's procurement area, the team confirmed that a multilevel sponsorship model existed and that it was the right sponsorship.

The program team then secured the highest level of sponsorship and commitment from the business service provider, as that had been lacking. The core team leveraged the program value justification (PVJ) technique to analyze the initial business case that supported the outsourcing via the eProcurement solution. The team revisited the business drivers, strategic objectives, outcomes, and expected benefits with the company's sponsors and then reengaged with the business service provider's sponsors to assess the chances of realization of benefits and the risks of a deployment failure. The program manager for this transformational change initiative led a current-state operational assessment in which the business service provider's model, service provider contracts, global implementation plans, and service level agreements were reviewed and analyzed.

BUSINESS OUTCOMES AND BENEFITS

The program manager of this strategic initiative identified and engaged the right sponsors, which facilitated strategic decision making. The PVJ and the operational assessment provided the sponsors with the specific information they were seeking. They deemed that the viability of the eProcurement solution smoothly taking over the company's existing requisition to payment process and systems was low. The analysis highlighted that the likely benefits of the outsourced model were not proportional to the risks being undertaken.

The company's sponsors signed off on the program management team's recommendation to cancel the multimillion dollar initiative on eProcurement. The go forward directive to the company's core team working on this transformation program was to evaluate other ways to innovate the procurement function. The sponsors shared their expectations of the business outcomes and business benefits from the procurement transformation and confirmed their support for the new direction of the program. The timely intervention by the sponsors mitigated the strategic risks and curtailed the losses from the proposed eProcurement initiative at this Fortune 50 high tech hardware conglomerate.

SUMMARY

The sponsorship team needs to own the transformation program and be visibly involved. The drivers for the business transformation need to be accepted and reinforced by the sponsors. The sponsors' alignment to the program strategy, vision, objectives, and outcomes needs to be evident to the program stakeholders. Explicit periodic communication of the sponsors' commitment to the transformation program will support the cause. A cross functional and multilevel sponsorship model is needed to authorize and legitimize the program.

The program's value justification technique provides the facts, data, and objective based business case for the program to keep marching ahead until the attainment of the program mission. The proactive, upward management of the sponsor by the program management team is key for the program team to get the continued needed support. Securing sponsor sign-offs on the critical change requests helps in managing sponsor expectations. By addressing the program progression barriers, the sponsorship team positions the program team to realize the desired business outcomes expected of the transformation program.

REFERENCE

PMI. 2013. *The sponsor as the face of organizational change.* White paper. Newtown Square, PA: Project Management Institute.

5

Develop and Implement a Governance Model

Programs designed to transform business are complex and usually impact multiple business functions. Such programs not only need the strong cross functional leadership to successfully lead the business change that is needed to get to the intended future state, but these programs also require a sound method for managing the program through the transformation life cycle. Transformation program governance sets up this monitoring method while also institutionalizing the strategic controls needed for the program to stay on course. Based on program business objectives and scope, some business units and functions may be less impacted than others by the program, but the cumulative impact of business change on the enterprise is high. If the program governance model is not put in place early enough, the ability of the program to manage the cumulative change impact is diminished, which poses substantial risk to business transformation.

As program managers are responsible for driving the transformation program forward, they must work closely with the executive sponsors in the early stages of the program to develop and implement the program governance model. The program management team needs to communicate the governance model to stakeholders, devise the governance processes and tools to support the governance model, and educate the program team to facilitate a successful roll-out of governance. The program manager can apply a combination of techniques in the program management tool kit to design and roll-out the governance model.

The following topics are described in this chapter with the help of supportive illustrations, including a real world case study:

- Governance modeling technique: Overview, objective, and approach
- Program communication and escalation protocol

- Program governance: Bodies, responsibilities, and rhythm
- Program accountability
- Governance policy design technique: Overview, objective, and approach
- Program governance: Decision making framework and change control management
- Governance of benefits realization
- Case study: Transformation program for postmerger integration

GOVERNANCE MODELING TECHNIQUE

Overview

A program governance model is a combination of governing and executive bodies, strategic control and oversight functions, and cohesive policies that define the consistent management of the program through the program life cycle. The five components of the governance model (Figure 5.1) are:

- Program organization structure to provide reporting clarity
- Responsibilities of the governing and leadership bodies to convey role clarity
- Determination of program accountability
- Escalation paths for issues and risks
- Program decision-making guidelines

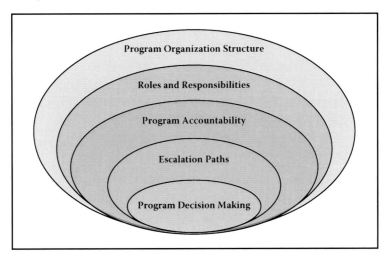

FIGURE 5.1
Governance model for transformation program.

Governance modeling is a set of techniques for consistently defining and recording the steps a transformation program team takes to realize the program vision and achieve program business objectives. The applied techniques are structured and flexible enough to meet the objectives of different kinds of programs, emphasizing pragmatism over dogmatism.

A governance model is not an organizational view, although executive and governance bodies often bear functional names. It is not a description of locations, although locations can be organized around the governance processes. A governance model is independent of technology, of core business processes as they currently happen, of existing organizational procedures, of the sequence in which events happen, and of the time at which they happen. In terms of deliverables, the governance model comprises multiple deliverables.

Objective

The goal of a well-designed and well-implemented governance model is to be a:

- Method to effectively manage the transformation program, which has so many moving parts
- Way to decide which organizational structure will most efficiently and effectively support the execution of essential processes, activities, and procedures constituting the program
- Vehicle to define the roles, responsibilities, and accountabilities
- Platform for creation, launch, and communication of governance body meetings
- System for identifying the escalation paths necessary to quickly eliminate the strategic barriers encountered by the program in realizing the program mission
- Framework for making various decisions as the program progresses

Approach

The development and roll-out of a governance model that supports a business transformation program entails the following steps:

- Understand and analyze the organization structure
- Confirm program scope and alignment to organization structure
- Build a straw program governance model
- Finalize the program governance model
- Communicate the program governance model

An elaboration of the above five steps embodying the governance modeling technique follows and this will facilitate an effective application of this technique by the business transformation program core team in order to create the needed artifacts.

Step 1. Understand and Analyze the Organization Structure

The best way to start building the various components of the governance model is gathering and analyzing information on the organization, the reporting hierarchy, the stakeholders, and the roles and responsibilities. A combination of insights and inputs gathered through one-on-one meetings and small group facilitated sessions with executives can be used to start formulating ideas and options around the program structure.

Step 2. Confirm Program Scope and Alignment to Organization Structure

The scope of the governance model should be consistent with the scope of the transformation program as defined at the starting point in the program charter. Do not get sidetracked in gathering and reviewing information on prior program organizational structures whose charter and scope boundaries were either too narrow or too broad. It is important for the governance model to factor any external environmental factors (outside business partners and service providers, outside processes, contractual terms, etc.) that might be germane and codevelop governance with external entities to ensure alignment.

Step 3. Build a Straw Program Governance Model

It can be useful to start with a straw model. The straw model is a rough approximation of the governance model used to encourage discussion and analysis. Straw models are sometimes called "straw men" and are meant to be taken apart, examined, and rebuilt. The straw model may have been created on another transformation program or by an outside business partner. Use it to generate ideas quickly, but beware of the problems it can cause:

- Leads participants to assume the model is correct
- Minimizes creative thinking by mimicking program structure to align to functional structure

- Possibility that program stakeholders will not examine the model as carefully as they should
- Discourages ownership of the model because participants do not contribute to its creation

Step 4. Finalize the Program Governance Model

This step is the process of refinement of the straw-man model to accommodate business changes, new information, and stakeholder inputs. The straw model is finalized over a series of iterations, and final sign off is secured at the highest leadership level involved on the program. This approach results in securing the buy-in of key decision makers and stakeholders.

Step 5. Communicate the Program Governance Model

This last (but not the least!) step entails sharing the model with the entire program team and all program stakeholders. The best practice approach is to develop and implement a communication plan that leverages the existing communication vehicles.

Sample Program Governance Model

The program management discipline is instrumental in facilitating business transformation due to its ability to understand, integrate, and manage the six dimensions of strategy, people, process, technology, structure, and measurement. Figure 5.2 presents a best practice governance model, with each dimension being composed of one or more tracks. For example, the process tracks may have multiple project managers with core teams beneath if multiple functional areas fall under the charter of the business improvements the transformation program is driving. In terms of another example, the people tracks may have multiple projects beneath, with each addressing a distinct organizational change area, i.e., change definition and impact, training, communications. Track sponsors own the tracks, with senior project manager(s) overseeing the multiple project managers leading the work within that track with the help of the core project team.

The senior project manager(s) get their direction and guidance from the program manager and they report to the program manager. If the program necessitates substantial program management capability, the

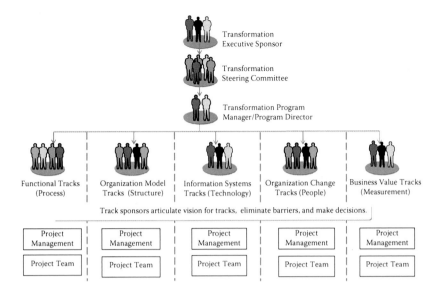

FIGURE 5.2
Sample program governance model.

organization's governance model will have multiple program managers, and they would report to the lead program manager. Depending on the scale of the transformation program and on how the enterprise is organized, one could have multiple program managers even at the track level. Due to the complexity inherent in transformation programs, it is not uncommon to have multiple lead program mangers and that group rolling into a program director or program head.

The program director (or an equivalent level in the structure) reports to the program steering committee, which is the senior leadership team representing the business functions and has the authority and accountability for those business areas. If any strategic external business partners are heavily involved in managing the transformation program, they also are represented at the steering committee level. The steering committee reports to and is accountable to the executive sponsor of the program. The executive sponsor (or sponsors) of the transformation program is the highest level executive with the responsibility, authority, and accountability for the program to achieve business objectives and realize the targeted business outcomes.

PROGRAM COMMUNICATION AND ESCALATION PROTOCOL

The program governance model drives transparency and facilitates streamlining of program communication and escalation of issues, risks, and other program matters. As the number of individuals working on a transformation program can be very large, and since they may be coming from different parts of the enterprise, high level guidelines around communications and specifications for escalation protocols will result in effective communications and escalation activities. Typically, the communication pathways align to the program's organizational structure. For example, project managers will issue formal communications about their respective projects to the program managers and not the steering committee. This protocol holds true for escalation as well.

Another typical example is that formal communications of the program (status, issues, risks, change requests) are delivered by the program manager to the steering committee and not to the executive sponsor, and this protocol applies as well for escalation of pertinent issues and risks. The intent of the communication protocol is not to create a rigid or a closed communications environment, but to have simplified and consistent messaging to program stakeholders to minimize potential confusion. Escalated items are typically complex in nature, necessitating deeper analysis and needing cross functional involvement to resolve. Swift resolution of escalated items is critical to minimize any negative impact on the transformation program, and the presence of a predefined escalation protocol will help the program team to resolve any escalation issues.

PROGRAM GOVERNANCE: BODIES, RESPONSIBILITIES, AND RHYTHM

The governance bodies can be planned and implemented on the foundational component of the organization governance model. There are two pieces: One is the formation and launch of the various governance bodies that meet regularly to oversee and direct the program, and the other defines the roles and responsibilities of each of these governance bodies. The cascade of leadership level meetings of these governance bodies is the governance rhythm, and the attendees in these forums could evolve a bit due to changes in the human resource allocated to the program.

The Boston Consulting Group's (2013) research paper on *Strategic Initiative Management* spells out the following as one of the four imperatives for executing strategic initiatives: "Establish program level routines that track milestones and objectives, communicate progress, and help identify issues early." The definition and communication of roles and responsibilities along with the charter for these ongoing governance meeting forums streamlines the agenda, avoids overlaps, and drives timely actions. Figure 5.3 aligns to the sample program governance model presented in this chapter.

Forum Name	Objectives of Forum	Forum Frequency	Forum Leader	Forum Attendees
Executive Sponsorship Meeting	Make strategic decisions on program direction. Eliminate road blocks. Ensure alignment to strategy. Approve large funding requests.	Monthly to Bi-Weekly		
Steering Committee Meeting	Determine priorities. Make program decisions. Address escalations. Approve funding requests and resource additions.	Bi-Weekly		
Program Management Meeting	Share program progress. Make program and project decisions. Review program level issues, risks, changes, and dependencies. Address escalations.	Weekly		
Track Sponsorship Meeting	Facilitate cross track collaboration. Review cross track matters. Sign-off on major changes to the track. Make resources available for each track.	Bi-Weekly		
Project Management Meeting	Report project level progress. Make project level decisions. Review project level issues, risks, changes, and project deliverables. Address escalations.	Weekly		
Core Team Meeting	Review select deliverables. Report activity level progress. Discuss raised issues and risks. Share ideas, learning, and best practices.	Weekly		

FIGURE 5.3

Transformation program governance rhythm.

PROGRAM ACCOUNTABILITY

The program governance model drives accountability at multiple levels. The development of program organizational structure, definition of governance bodies, and delineation of roles and responsibilities assists in providing clarity at an individual, team, and governance body level. The governance model ensures that the authority is appropriately granted and understood. The definition of individual roles (i.e., project manager, program manager, steering committee member, etc.) assists with understanding the individual level accountability, with the executive sponsor having the ultimate accountability for realization of the program's desired business outcomes. Those who are part of a team are accountable for the outcomes, work products, and results expected of that team. The members of a governance body are collectively accountable for delivering against the charter of that forum.

GOVERNANCE POLICY DESIGN TECHNIQUE

Overview

The building blocks of a solid governance model are cohesive policies that enforce the consistent management of the transformation program. A program governance policy is a parameter or rule regarding a specific action or issue that dictates how the action on the program should be performed or how the program issue should be resolved. The program governance policy provides direction when executing a program process that has several possible approaches. Policies help direct the actions of the program team and should not be so restrictive that it hinders the team's performance. Instead, they should help guide the program team and stakeholders in the right direction. The program governance policy should be clearly communicated to program stakeholders and should not conflict with program value statements.

Objective

In designing policies for transformation program governance, the primary goal is to prevent undesirable business outcomes. It is important to focus the development of policies on managing rather than on controlling program management processes. However, it is often necessary to include control points. The purpose of these governance policies is to develop a set of guidelines the program team can follow in order to diligently and effectively execute the program management processes. These program governance policies should not be rigid edicts that can hinder the transformation program team from efficiently accomplishing the program's business objectives.

Approach

Designing a program governance policy involves finding a source of policy, drafting that policy in appropriate detail, and finally ensuring that the policy is consistent with other policies and with the transformation program's value statements. It entails the followng steps:

Step 1: Review Program Management Processes and Role and Responsibility Descriptions

The first step in designing program governance policies is to look for the potential sources of policies. The current program management processes, business organization structure, reporting hierarchy, and roles and responsibilities need to be studied. The previously identified constraints for potential sources of governance policies should be revisited. Some of the program management processes and newly established program governance bodies will need to be supported by new policies. Process driven policies can often be found by examining decision points. Decision points often capture implicit policies, but policy design makes them explicit.

Step 2: Draft Appropriate Program Governance Policies

After identifying the sources of policies, draft the governance policies based on the level of detail required. These policies should be developed in synchronization with the program management processes that will be affected, since a policy may have a direct impact on the way a program management process is performed.

Step 3: Add Control and Audit Points

Determine and document all the control and audit points identified while tracing all the activities in a program management process. In identifying and documenting these points, keep in mind that the purpose is not to introduce a rigid control structure but to develop a guideline for managing the program management process consistently and effectively.

Step 4: Ensure Program Governance Policy Consistency against Program Value Statements

Once the governance policies have been drafted, they must be cross verified against one another to make sure that there are no conflicts between them. A new governance policy that is drafted for one program management process may violate an existing policy pertaining to the same or another program management process. Not only must these policies be cross verified against one another, but they also must be checked against the program value statements.

Step 5: Clearly Communicate the Purpose of the Governance Policy to Program Stakeholders

Once the program governance policies have been drafted and approved, the policies and their objectives should be clearly communicated to the program stakeholders. The communication should highlight the significance of the adopted policy and what it will accomplish for the transformation program. Such a communication minimizes potential confusion among the program team and stakeholders and makes it more likely that these governance policies will be supported and followed.

PROGRAM GOVERNANCE

Decision Making Framework

One of the components of the program governance model is the high-level definition of the decision making processes. This definition is a critical element, as many decisions will have to be made during the execution of the program life cycle. From a strategic program control perspective, an example is the key "go/no" decision that needs to be made at critical junctures or "stage gates" of the transformation program. If a transformation program is led with the program manager making all project level decisions and the project manager making all program level decisions, the probability of failure of business transformation is high. The framework for decision making draws a distinction between program level and project level decisions in terms of the decision maker. The framework also accounts for the different steps needed to make those decisions.

As the program management team drives the transformation program forward, key examples of decisions (program and/or project level) to be made may pertain to the constraint areas of scope, timeline, budget, quality, risk, resources or value. Other types of decisions to be made could be in the areas of prioritization, change request, or program dashboard status. The impact a decision would have on the transformation program will determine the decision maker, the urgency of decision, and the extent of due diligence needed prior to making the decision. In addition to the impact factor, the other important factor governing the decision making process is the severity of the issue that has surfaced and/or the level of a known risk to the program. A grid that maps these two factors (business impact and issue/risk level) not only accounts for them, but also drives the criteria for escalation of decision making (Figure 5.4). The framework resulting from the two dimensional grid determines the optimal decision maker, and this could be the project manager, the program manager, the steering committee, or the executive sponsor.

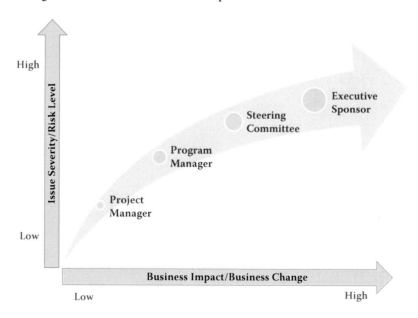

FIGURE 5.4
Program governance: framework for decision making.

Change Control Management

The design, roll-out, and implementation of change-control processes for a transformation program are important activities. External market forces, internal developments, and new revelations as program execution progresses can trigger change requests. As there will be a spectrum of program level and project level change requests that impact the program in various ways, the critical ones are those directly impacting the delivery of planned program business outcomes. The program manager should determine whether or not to approve, defer, or reject the change request; he or she escalates it to governance as needed and as stated in the program governance model. The framework in Figure 5.5 is used to capture, analyze, and document decisions on such change requests.

As the key value indicators (KVIs) and value metrics are used in measuring business outcomes, the program management will review them along with the program health assessment criteria—timeline, funding, resources, scope, quality, risk, and value—in assessing the impact of the change requests on the transformation program. To facilitate effective and timely decisions on such change requests, the program manager(s) or program management has to involve the right subject matter experts and senior management in the decision making process. The program manager can secure inputs of the steering committee and executive sponsor on any change requests that significantly impact the business outcomes expected by the stakeholders. The rigor in securing sign off on the change request decision, creating an appropriate plan, and then implementing associated communications helps in motivating the team tasked with the business transformation program.

Id	Change Requestor	Change Request Description	Business Outcome Impact	Timeline Impact	Funding Impact	Resources Impact	Scope Impact	Quality Impact	Risk Impact	Value Impact	Decision and Rationale	Sign Off
1												
2												
3												

FIGURE 5.5
Sponsor involvement in approval of critical change requests.

GOVERNANCE OF BENEFITS REALIZATION

The preceding sections touched upon the importance of rigor in the change control management and decision making processes. As the program management team furthers the transformation program by executing the transformation program plan, it is possible that certain change requests could have an impact on the prior agreement and strategic assumptions pertaining to business objective attainment, business outcome delivery, and benefits realization. The development and roll-out of the benefit realization governance process formalizes the recalibration of desired business outcomes and expected business benefits. As part of the analysis done on a change request, the impact on expected benefits from a transformation program needs to be understood, documented, and communicated to the stakeholders. The program manager has to take the steps needed to secure buy-in of the pertinent stakeholders and approval of sponsors to the changes in the realization of the business benefits and delivery of business outcome.

The benefits realization governance can also validate that the program business outcomes have been well defined and follow the SMART principle. A SMART business outcome is one that is specific, measurable, attainable, relevant, and time based. *Specificity* provides clarity while also eliminating overlapping business outcomes. The need to pass the litmus test of being *measurable* is a critical one. *Attainable* refers to a business outcome that is a bit ambitious, but definitely feasible to accomplish. *Relevancy* enables tracing back to the program vision alignment to the program scope boundaries. The business outcome has to be realized within a designated *time frame.*

The SMART acronym is an easy way of remembering the guiding principles that aid in crisp and clear definition of desired program business outcomes. The initial governance around the definition and agreement of program business outcomes and program benefits is important. Also, the ongoing governance around any changes to the definitions and agreements of outcomes and benefits mitigates the risk of the delivered outcomes being misaligned with expectations.

Case Study: Transformation Program for Postmerger Integration

CONTEXT

One of the largest biotechnology organizations with global operations had acquired another leading biotechnology company to further dominate the global market. The acquired entity was integrating operational functions, consolidating facilities, and migrating information systems.

BUSINESS PROBLEM OR OPPORTUNITY

One of the most critical integration areas for the biotech company was the contract to cash operational function and the supporting systems. If the contracts were not properly streamlined and integrated, the large customers of the acquired biotech company would be negatively impacted. Slow decision making was crippling the integration and roll-out of the streamlined contract to cash business process. The points of accountability and the roles and responsibilities of the numerous personnel from the various functions involved in the contract to cash process were not defined. As the integration timeline was dynamic and the integration had to comply with the regulatory guidelines, the lack of a change control process on the multiple projects integrating the contract to cash operations was a significant risk to the postmerger integration program.

SOLUTION

The contract to cash merger integration program team stood up a merger integration office to organize, track, and coordinate the various projects needed to complete the merger initiated transformation of the contract to cash operational function. In the course of developing the structure of the integration office, the program governance model was developed and points of accountability established. The roles and responsibilities of the governance bodies were defined, and a program rhythm that would enable rapid decision making and timely resolution of escalated issues was implemented.

The integration office program manager created, socialized, communicated, and rolled out processes for program escalation, compliance management, and change control management. The core program team comprising subject matter experts from both the biotech companies collaborated in developing an integration road map. The team dedicated to contract to cash integration had to go through the change control governance body if any changes to the signed-off and baselined integration road map and project plans were needed.

BUSINESS OUTCOMES AND BENEFITS

The implementation of the merger integration program governance rhythm institutionalized the needed controls, enabled timely decisions, and mitigated risks. The benefits realization governance process ensured that the regulatory compliance requirements were baked in as part of the merger integration activities happening within the acquiring biotech company and the acquired biotech company. The aggressive timeline to attain the future state of an integrated contract to cash process was realized. The design and launch of the governance model by the contract to cash merger integration office played a key role in the attainment of the merged future state.

SUMMARY

For a transformation program to succeed, it is imperative to develop and implement the governance practices sooner rather than later. The program governance model is a combination of governing bodies, strategic control and oversight functions, and cohesive policies that defines the consistent management of the program throughout the program life cycle. The key constituents of the program governance model are: program organization structure, governing body definition, roll-out of governance forums, program accountability, escalation paths, and decision making processes. Program governance ensures that strategic direction and program vision are aligned, that program priorities are defined and understood, and that decisions are aligned with the overall business objectives. A lack of robust governance practices poses a substantial risk to realizing the desired business outcomes targeted by the transformation program.

REFERENCE

Boston Consulting Group. 2013. *Strategic initiative management: The PMO imperative.* Boston: Boston Consulting Group.

6

Define Success, Outcomes, and Key Value Indicators

What defines success for a program? What business outcomes does the program need to realize? What are the key value indicators (KVIs)? It is important that these questions be answered at the onset of a transformation program initiation. Similar to business executives keeping a close quarterly tab on their company's top and bottom line performance, a program manager needs to continuously keep track and report progress of the initiative through the agreed upon key value indicators. Chapter 2 discussed the "must have" subject matter content in the program charter deliverable, and one of those is the clear articulation of what constitutes program success.

The business outcomes being targeted and expected by stakeholders have to be established. Key value indicators that would validate the realization (or lack thereof) of business outcomes need to be identified. High due diligence and precision are necessary in determining the value metrics supporting the key value indicators. In Chapter 3, the program management life cycle detailed in the book was mapped to the Project Management Institute's (PMI) third edition of the *Standard for Program Management* (2013a). Some of the topics expounded in this chapter pertain to the benefits management domain in the PMI standard.

The following topics are described in this chapter with the help of supportive illustrations, including a real world case study:

- Performance improvement measurement technique: Overview, objective, approach, and helpful tips
- Positioning transformation program to deliver business outcomes
- Business outcome modeling technique: Overview, objective, and approach
- Case study: Selection program for enterprise system

PERFORMANCE IMPROVEMENT MEASUREMENT TECHNIQUE

Overview

Performance improvement measurement is a technique to measure and evaluate the improvement in business performance as a result of successful completion of a transformation program. In other words, application of this technique will determine whether the program delivered the intended business benefits at program closure. Performance improvement measurement can also assess whether the benefits delivery is being sustained after the program has been transitioned for ongoing execution by the operations function. Many enterprises have successfully developed and implemented an improvement measurement architecture concept, which is a comprehensive approach to address measurement, the sixth dimension of program management.

Performance improvement measurement architecture is an integrated and balanced business performance measurement system supporting the transformed enterprise. It is a top down program performance measurement system, unlike most legacy measurement systems, which have tended to evolve from the bottom up. Using a top down decomposition approach, the agreed business objectives are translated to business outcomes. The high level key value indicators (KVIs) for each business outcome are determined, and the value metrics supporting the KVIs are defined. KVIs focus on the "business value" created by the transformation program, which can be tangible or intangible business results, business benefits, or business outcomes. The KVIs are higher level determinants of business performance, and value metrics are the lower level measures of business performance. The delivery of business results, benefits, or outcomes by a business transformation program is ascertained by measuring the level of business performance in the attained future business state with the help of the agreed KVIs and value metrics.

Objective

Performance improvement measurement provides an enterprise with:

- A comprehensive framework to translate the transformation program mission, vision, strategies, stakeholder values, and business objectives into an integrated set of business performance assessment measures that can be linked throughout the organization

- The ability to balance external KVIs and value metrics with internally focused KVIs and value metrics necessary to effectively assess the success of the transformation program
- The ability to determine how higher quality, lower cost, and reduced process cycle times translate into improved business outcomes like higher sales, more market share, reduced operational expenses, improved share prices, and higher return on assets
- A performance improvement measurement dashboard grounded in the strategic business objectives and is reliant on a concise group of KVIs
- Greater transparency to the stakeholders of the transformation program
- The ability of every individual to understand how the program and how the individual add value to the areas (enterprise, division, function, process) part of the transformation program

Factors

The following factors should be taken into consideration when developing the performance improvement measurement architecture:

- For each program objective, only the critical key value indicators are decided upon, as having too many key value indicators creates complexity and room for inconsistency.
- Financial KVIs are articulated to indicate how the program strategy and the other four program dimensions (people, process, technology, and structure) are contributing to the bottom line results.
- As the KVIs and value metrics represent a true measurement of performance, they need to be able to evaluate the realization of the short term and long term business outcomes.
- Management needs to be able to see the impact an improvement or investment in one area will have on the other areas.

Approach

The process for implementing performance improvement measurement technique entails the following eight steps:

- Ensure the program business objectives have been articulated
- Confirm the scope

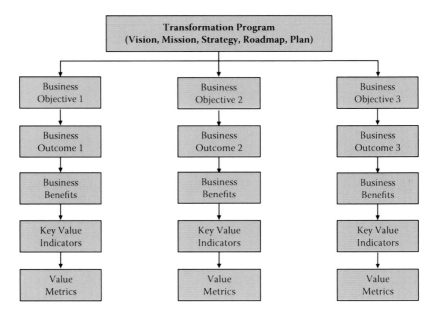

FIGURE 6.1
Framework for assessing transformation program success.

- Define the business outcomes
- Develop the first cut improvement measurement architecture
- Define the business benefits
- Identify the KVIs
- Determine the value metrics
- Set targets for value metrics and map to business process

A detailed description of the above steps composing the performance improvement measurement technique follows as it will enable its effective application by the business transformation program core team in order to develop the needed deliverables (see Figure 6.1).

Step 1: Ensure the Program Business Objectives Have Been Articulated

These are the statements of purpose for the program's existence and represent the highest, long term business goals of the enterprise. The best practice is to have objectives for each of the major categories of applicable stakeholders—customers, employees, operations, and shareholders. The highest level long term goals are captured in the form of business objectives. Each high level business objective is decomposed into the lower level business outcome. The KVIs for each business outcome are determined, and value metrics supporting the KVIs are identified. For example, a customer related objective will specify what value the company will provide to the customer, and that should address a specific customer pain point or an unmet key expectation.

Step 2: Confirm the Scope

The scope for the development of the performance improvement measurement architecture must align to the scope of the transformation program.

Step 3: Define the Business Outcomes

The body of work done to align transformation program strategy with the operations and technology strategies serves as the key input to creation of program business objectives. These business objectives can be drilled down into business outcomes, which need to be finalized after socialization with pertinent stakeholders.

Step 4: Develop the First Cut Improvement Measurement Architecture

Facilitated sessions are recommended for this step. Given the complexity of a business transformation program and the significant business changes such a program is driving on multiple fronts, a holistic performance improvement measurement framework is warranted. At a minimum, the perspectives of the following stakeholders need to be incorporated:

- Customers
- Employees
- Operations
- Shareholders

Step 5: Define the Business Benefits

Chapter 2 covered the business outcome delivery strategy and the benefits realization definition. The upfront clarity in defining the business benefits sets the program up for success. The defined benefits should be validated as the benefits realization measurement infrastructure is being put in place.

Step 6: Identify the KVIs

Through facilitated sessions, each business objective is examined in turn, and a list of KVIs is brainstormed. It is recommended that there be no more than three KVIs for each objective.

Step 7: Determine the Value Metrics

The KVIs have to be analyzed to derive the value metrics that aid in performance improvement measurement. Though the program manager is unlikely to be the subject matter expert, the program manager ensures that relevant experts are engaged in the analysis sessions. The defined value metrics should provide the performance improvement information necessary to support the strategies and mission of the transformation program. The initial longer list of value metrics is refined and prioritized to restrict it to three value metrics per KVI. The following list presents a couple of examples by stakeholder group:

- *Shareholder*: Metrics that reflect the shareholder values should include financial operating measures that can be used to run the everyday business. An example of a financial related KVI is revenue growth. Examples of the value metrics aiding the revenue growth performance improvement measure include revenue growth by product, by geographic region, by quarter.
- *Customer*: In developing customer related value metrics, it is important to factor the tiers or segments of the customer base. A relevant KVI example pertaining to customers would be *customer satisfaction*. Examples of value metrics to gauge performance improvement on the customer satisfaction KVI include the number of referrals from customers, movement in the customer satisfaction index, number of customer complaints.

Step 8: Set Targets for Value Metrics and Map to Business Process

For each identified value metric, performance improvement targets have to be set. The key considerations in coming up with targets are as follows:

- *Mapping of the metrics to the organization area, business function, and process.* This mapping ensures that it is known which organization, function, and process is responsible for meeting the performance improvement targets.

- *Determining how the measurements impact each other.* For example, a reduction in a cycle time metric of the order fulfillment cycle is likely to have an impact on a cost metric for that cycle.
- *Understanding of how these metrics are to be reported* and what the characteristics of a supporting information system are to be.

Helpful Tips

- Customer based business objectives, business outcomes, KVIs, and value metrics are important. These must be mapped to the business operations side, as the processes and their supporting infrastructure ultimately determine how well the customer needs can be met. The integrated transformation program plan needs to factor and manage the work that needs to be executed on the operations side.
- To achieve cycle time, quality, and cost objectives of the transformation program, the outcomes, KVIs, and value metrics should be decomposed to a level that can influence employees. This decomposition is important, as it ensures that the performance improvement indicators are linked to the actions taken by individuals and addresses the "What is in it for me?" question.
- It is important to realize that as each industry, market segment, and company is unique, so are the key value indicators for each.
- One critical success factor in building and implementing the integrated performance improvement measurement architecture is that the company's information systems be able to support the timely analysis of the KVIs and metrics. If the information systems cannot support such an analysis, much of the value to be obtained from the architecture cannot be realized.

POSITIONING TRANSFORMATION PROGRAM TO DELIVER BUSINESS OUTCOMES

A business (or technology) transformation program is made up of multiple projects. It can also include other programs (or subprograms) based on the magnitude and complexity of the transformation program. The program management team uses the measurement architecture framework to get a good understanding of the linkages from a measurement point of view.

The typical transformation program organizational structure would have multiple project managers driving the projects assigned to them. It is common to have multiple program managers too, which results in a couple of clusters of program teams. The expected short term and long term business outcomes need to be tied to the program organization structure.

The concept and development of the transformation program road map was discussed in Chapter 1 at a high level. The program business objectives, assessment of the organization capabilities, and business architecture work determine the "business capabilities" the transformation program has to deliver. The business capabilities are the highest level roll-ups of the requirements or desired functionality. The desired business outcomes expected from a transformation program are derived from the program business objectives. The focus, as well as the visibility into how the transformation program will deliver the expected business outcomes, can be best established and communicated through the following three maps:

- Business capabilities to business outcomes
- Business capabilities to projects
- Business outcomes to timeline

Figure 6.2 portrays the three maps with the help of a program comprising three project over a four year time horizon. The program has to deliver four business capabilities and realize nine business outcomes.

BUSINESS OUTCOME MODELING TECHNIQUE

Overview

The program management team has to orchestrate the process of setting clear, measurable, and achievable business outcomes for the program. Business outcome modeling defines and prioritizes the business outcomes that a transformation program is chartered to deliver. As seen in Chapter 2, one of the process steps in designing the performance improvement measurement architecture is the definition of business outcomes. Ideally, the business outcomes are linked from the top of the enterprise to the bottom, so that employees can see how they directly contribute to the success of the enterprise.

	Business Outcomes To Be Realized
Business Capabilties To Be Delivered	
A	1, 2, 5
B	3, 8
C	6, 7
D	4, 9

Management Capability to Business Outcomes

Business Capabilities To Be Delivered	Project 1	Project 2	Project 3
A	x		x
B	x		x
C		x	
D		x	x

Business Capabilities to Projects

Business Capabilities To Be Delivered	2014	2015	2016	2017	Total
Business Outcomes (Examples)					
1. Incremental Revenue					
2. Protected Revenue		All $ in	(Millions)		
3. Cost Savings					
4. Improved Profitability					
5. Higher Market Share					
6. Better Competitive Position					
7. Experience Enhancements					
8. Productivity Improvements					
9. Cycle Time Reduction					

Business Outcome Realization Timeline

FIGURE 6.2
Positioning program to deliver business outcomes.

Objective

Business outcome modeling can be used to:

- Focus the program to achieve business results in the most efficient and effective manner
- Create an inventory of business outcomes and prioritize them
- Develop a strong commitment, sponsorship, and ownership of the business transformation program
- Validate the strategies driving the direction of the program
- Identify the problems inhibiting the business processes from reaching tactical and strategic business outcomes
- Assess business issues or problems that inhibit the achievement of business outcomes

Per PMI's *Pulse of the Profession: The High Cost of Low Performance* (2013b) research study, organizations lose an average US$149 million for every US$1 billion spent on strategic initiatives due to poor program and project performance. The value derived from utilization of business-outcome modeling is tremendous, as it can help an organization avoid poor program performance.

Approach

Enterprises with higher success rates on their transformation programs and higher program management maturity do not assess business outcomes only at the start and the end of the program. Unfortunately, many others take the high risk approach of identifying desired business outcomes at the start and checking only at the end to see if they were realized. The best practice is to approach business outcomes as a continuum and adopt a life cycle approach to managing them. The business outcome life cycle management approach can be woven into the transformation program management life cycle. Program management can drive the end-to-end process beginning with business outcome definition and ending with business outcome sustainment. Figure 6.3 illustrates the best practice approach to business-outcome management on a complex business transformation program.

The following four stages of business outcome life cycle management are mapped to the highest level program management life-cycle processes:

- Outcome definition and validation
- Outcome planning and measurement mechanism
- Outcome delivery tracking
- Outcome sustainment and measurement

Stage 1: Outcome Definition and Validation

In Chapter 2, the definition and significance of business outcome delivery was elaborated. The outcome expected to be delivered by the business transformation program is categorized into these four buckets—customer, operational, employee, and financial. Figure 2.5 mapped the high-level process to be followed by the core program team to make the transformation program a business success in the eyes of the stakeholders. The program charter development process and the resulting signed-off program charter deliverable aids in defining the outcome expected from the program. The expectations of stakeholders around the outcome have to be validated early on and throughout the program execution life cycle.

Stage 2: Outcome Planning and Measurement Mechanism

The program charter artifact is a key input to the business transformation program planning process. The end-to-end program plan is one of the main outputs resulting from the planning process. The program plan ensures all the work, deliverables, and outcomes to be realized are clearly articulated along with the activities, resources, timeline, and interdependencies. The program plan supporting the business transformation is comprehensive as it has to capture and integrate the activities associated with all the six dimensions of program management—strategy, people, process, technology, structure, and measurement. Typically, a transformation program is structured into multiple phases with business outcome to be delivered at periodic intervals either during the phases or at the end of each phase. The plan spells out the measurement strategy and approach to be used to assess the delivery of the business outcome. The key stakeholders of the program sign off on the program plan and the nature and timing of delivery of business outcome.

Stage 3: Outcome Delivery and Tracking

The business outcome modeling technique adds rigor to the process driving the realization of outcome. The "make it happen" phase of the program management life cycle is the execution of the detailed, integrated program plan. As the plan is getting executed for the program to attain the desired future state, the monitoring of the program execution is happening concurrently. The program monitoring function assures the timely and proper completion of the planned deliverables and business outcome. The implemented measurement approach confirms the delivery of outcome specified in the program charter and the program plan. Any deviations in delivering the agreed upon outcome are addressed by the program management team. As the program plan gets refined and/or rebaselined

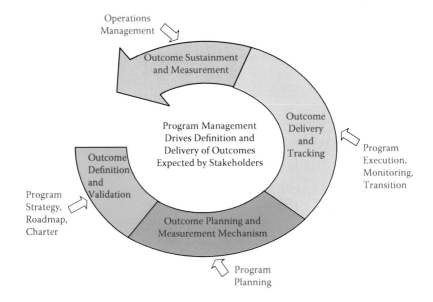

FIGURE 6.3
Business outcome life cycle management.

to account for any outcome delivery deviations, buy-in is secured through stakeholder engagement and leadership approval is obtained via the established program governance processes.

Stage 4: Outcome Sustainment and Measurement

The business objective of the final phase of the transformation program management cycle is the smooth transitioning to the desired future state and sustaining the expected outcome in the attained future state over the long haul. The business outcome modeling technique facilitates the sustainment of the intended outcome which is the ultimate gauge of success of the transformation program. The end-to-end business transformation program plan incorporates the activities needed to sustain the defined outcome and validate the continued realization of the outcome via the measurement mechanisms in place. The last stage of the business outcome life cycle management provides the oversight and support to ensure the operational functions are positioned to sustain the business outcome that was planned at the launch of the transformation program.

Case Study: Selection Program for Enterprise System

CONTEXT

A highly recognized financial capital management company that offers open ended mutual fund products to retail customers was experiencing exponential growth in the assets managed and customers served as a result of stellar performance of all six of the mutual funds it offers. The company's vision is to maintain a competitive edge by offering close ended mutual fund products to another segment of retail customers within a year. The company has also announced that it will start offering financial products and advisory services to the institutional market within two years. The company needs to be ready to offer these new products and services.

BUSINESS PROBLEM OR OPPORTUNITY

The company's systems are not capable of handling close ended mutual fund products. The open ended mutual funds cannot be traded in real time during the market hours, whereas close ended mutual funds can be, and they tend to attract a different set of retail investors. As the business processes, information systems, and human capital needed to take care of the institutional customer are different, the financial capital management company has to put that infrastructure in place to realize the communicated vision.

SOLUTION

As the strategic direction of this privately owned financial capital management company and the time horizons are clear, the program manager and the core team initiated a program to identify, evaluate, select, and implement an enterprise level system with the needed capabilities. The desired future state is an implemented enterprise system, redesigned processes, and a modified organizational model. The program management team facilitated a series of business outcome modeling workshops to define and validate business objectives, business outcomes, and business benefits that the chosen implemented system needs to deliver. The information needed to construct the performance improvement measurement architecture was gathered through the subject matter experts attending these workshops.

The architecture would be used not only to measure the business performance of the capital management company in the attained future state, but to also use it in the front end of evaluating and selecting the system that meets the defined business and technology capabilities and requirements. The program team managed the end-to-end process of system selection, which included issuing request for proposals to short-listed software vendors, evaluating proposal responses, reviewing product functionality in demos, conducting evaluations, finalizing selection with stakeholder inputs, negotiating with the preferred vendor, and formally awarding the contract to the software vendor. With the vendor's expertise, the conceived program plan for system implementation met the announced timelines for new product roll-outs.

BUSINESS OUTCOMES AND BENEFITS

The structured process adopted by the program team resulted in the selection of a solution that met the strategic objectives and stakeholder needs. The program plan not only had the systems implementation component plan, but also the component plans for the business process and organization resource buildup work. The performance improvement measurement architecture will assist with measuring realization of benefits upon successful implementation of the system, business process, and new organization. The financial capital management company's rigorous process in defining success, outcomes, and key value indicators during the early stages of the program and involving stakeholders in that process provided the platform for program success.

SUMMARY

The program manager has to drive and build consensus on the definition of program success, the business outcomes, and the key value indicators (KVIs). These definitions need to be revisited at key milestones throughout the program life cycle to confirm continued alignment. Such a proactive approach is similar to the construct of program risks needing to be identified, mitigated, and watched from program initiation through program closure. The importance of driving adequate clarity upfront on what constitutes the success of a transformation program is high.

Elaboration of the expected outcomes and timing the delivery of those outcomes are critical factors in the success of the program. Equally important is the periodic communication and education around the definition of a successful transformation program, targeted business outcomes, performance improvement measurement architecture, and KVIs to all stakeholder groups. Program management has to constantly manage stakeholder expectations and validate that there are no deviations from the original definition of success, the strategic criteria to judge success, and the KVIs. The activities associated with management of the business outcome life cycle have to be planned in the integrated transformation program plan, and these activities have to be executed and monitored.

REFERENCES

PMI. 2013a. *The standard for program management.* 3rd ed. Newton Square, PA: Project Management Institute.

PMI. 2013b. *The high cost of low performance.* Pulse of the profession in-depth report. Newton Square, PA: Project Management Institute.

7

Invest in Planning and Creating an Integrated Approach

Given the long life cycle of the typical transformation program, the itch to commence program execution without adequate planning is high. Program managers must resist the temptation to shortchange the planning process. The core foundation for a transformation program is its alignment to the strategic direction of the organization, which is accomplished through the program vision work described in Chapter 3. The layer over the core foundation is the program architecture, which results in a blueprint for the desired future state and business case for the program. Upon completing these two layers and securing the executive sponsorship to move forward, the program management life cycle commences. The program strategy is formulated, and a program road map to implement this strategy is developed as part of the front end of the program management life cycle.

The creation of an integrated, holistic approach to solve the defined business problem is imperative. The integration of the organization change management in the overall approach taken to transform is critical, as it effectively addresses the "people" dimension of program management. An integrated transformation program plan is next prepared to deliver the desired business outcomes specified in the signed-off program charter by the key stakeholders. All the steps taken so far constitute transformation program planning work, which provides the basis for program execution and program monitoring. Many transformation programs are unable to deliver on the promise due to the program planning being shortchanged. The investment of time, energy, and resources in planning for the business transformation program is mandatory for the program to transform the business, which results in the realization and sustainment of desired business outcomes.

The following topics are described in this chapter with the help of supportive illustrations, including a real world case study:

- Program architecture technique: Overview, objective, and approach
- Program architecture drives program management life cycle
- Organization change management technique: Overview, objective, approach, and drivers
- Organization change readiness
- Transformation program planning technique: Overview, objective, and approach
- Benefits realization planning
- Case study: Business transformation program to launch a new product

PROGRAM ARCHITECTURE TECHNIQUE

Overview

Program architecture brings together the business and technical architectures and creates the overarching platform for implementing enterprise strategy. Large enterprises frame up multiple strategic initiatives to get from the current state to the transformed future state from an enterprise level perspective. The breadth, depth, and scale of each strategic initiative can vary. The business objectives and business outcomes to be delivered in a certain time horizon could be through a single or a couple of strategic initiatives. The complexity of planning multiple strategic initiatives is high, which poses significant enterprise level risk. The risk can be mitigated through program architecture, another technique from the program management tool kit.

Objective

Program architecture establishes the bridge to implementation of strategy by providing a standard and unified platform for initiating and planning strategic initiatives through the program management discipline. Program architecture has visibility into all the initiatives on the enterprise level strategic initiative road map. The output of program architecture becomes the strategic input for the program management life cycle, the four phase approach to drive a business transformation program (or a strategic initiative) from start to finish. In the context of a designated business transformation program that has been initiated, the technique of program architecture analyzes the business problem thoroughly and designs the high level approach to solve the problem.

For example, if a business transformation program at a large enterprise entails radically redesigning major business processes and automating these processes, the program architecture will call out the need for a much greater focus of the program team toward the process and technology dimension. Program architecture facilitates adoption of a holistic approach that not only addresses all the root causes of the problem, but also sustains delivery of desired business outcomes even after the formal closeout of the transformation program.

Approach

The five-step process is depicted in Figure 7.1. This technique is utilized in the early stages of the business transformation journey to drive the *program discovery stage*, whose output is instrumental in providing the launch platform for the transformation program. All six of the program

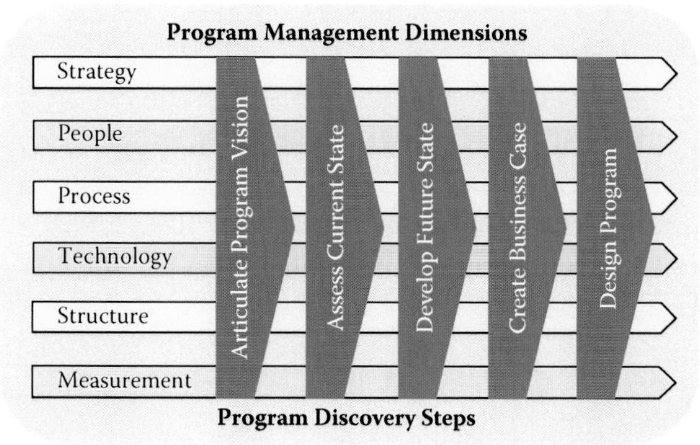

FIGURE 7.1
Program architecture: bridge to strategy implementation.

management dimensions of strategy, people, process, technology, structure, and measurement are taken into consideration in developing the program architecture.

1. Articulate program vision
2. Assess current state
3. Develop future state
4. Create business case
5. Design program

Step 1: Articulate Program Vision

In this step, the business transformation program's vision, mission, objectives, and stakeholder values are elaborated. Though these will be distinct for each program and an important key step for each program, they all align to the enterprise level vision, mission, objectives, and stakeholder values. As part of elaborating the strategic alignment technique in Chapter 3, the processes of create program vision, develop program mission, determine program stakeholder values, and develop program objectives was touched upon.

Step 2: Assess Current State

Assessment of the current state aids in deeper comprehension of the problem statement, the pain points, implications, and root causes. The magnitude of the problem and its complexity can be revealed in assessing the current state across all the six dimensions, i.e., studying the current processes and identifying gaps, understanding roles and skill sets of personnel involved in current processes, etc. The voice-of-customer and business performance calibration are sample techniques that can be put to use to get arms around the current state of the business pertaining to the scope of the program.

Step 3: Develop Future State

A high-level picture of the desired future state is created with the input of the subject matter experts, designated stakeholders, and the program management team. Though the steps are being covered linearly, some of the future state definition work is happening in parallel with the analysis of the current state. The options, solutions, and ideas to efficiently address the gaps identified during the current state assessment are covered as part of future state development. Again, all six of the program management dimensions have to be factored, e.g., changes in the technology environment to accommodate the automated future process or organizational structural changes to minimize handoffs from one group to the other to provide an improved experience. The environment scanning technique can provide valuable inputs on internal and external best practices for incorporation.

Step 4: Create Business Case

The techniques of program value justification and performance improvement measurement can be used to analyze the information gathered to date to develop a fact based business case model that justifies continuation and full launch of the transformation program. The business case has to showcase the program's ability to realize the program vision, achieve program business objectives, attain the desired future state, deliver the business outcomes, and sustain the outcomes. As the business case model is targeted at an executive sponsor, a steering committee, and other senior leaders, it has to effectively elicit the "program value" in their eyes.

Step 5: Design Program

The *design program* step drives the strategic approach to be taken by a program to achieve the program's business objectives. The crux of the approach in business problem solving resides in whether and how the program management dimensions of strategy, people, process, technology, structure, and measurement are accounted. No two business transformation programs are exactly the same, as each is solving a unique problem (or capitalizing on an opportunity) under a unique context. Program design output specifies which dimensions warrant a higher focus in the program management life cycle. For example, a large program with the charter of redesigning the organization is likely to have minimal need for a technology development effort. The design program step also ensures that the overlaps with other in-flight and yet-to-be-launched programs are being taken into consideration.

PROGRAM ARCHITECTURE DRIVES PROGRAM MANAGEMENT LIFE CYCLE

The first program management life-cycle process of "formulate program strategy" under the first phase of "set the stage" picks up from where the program architecture stops. As a recap, the program management life-cycle phases and the processes under each phase are listed here:

Phase 1: Set the stage
- Formulate a program strategy
- Develop a program road map

Phase 2: Decide what to do
- Define the program charter
- Create the program plan

Phase 3: Make it happen
- Execute the program
- Monitor the program delivery

Phase 4: Make it stick
- Transition to operations and close the program
- Sustain outcome delivery

Program strategy, program road map, and program charter drive the development of the transformation program plan, which is described later in this chapter.

ORGANIZATION CHANGE MANAGEMENT TECHNIQUE

Overview

Program management should embed the organization change management (OCM) process to strategically mitigate the risk of failure of a transformation program. One of the origins for lack of success in a business transformation effort is the failure to take into consideration major aspects of strategic change enablement. Poor organization change readiness to counter the business impact, inadequate stakeholder engagement, ineffective communications, and lack of timely training are potential causes of lack of acceptance of the business change. The failure to embrace change results in nonachievement of intended business objectives and nonrealization of desired business outcomes.

The broader and deeper the impact of a complex program, the greater the criticality of addressing the "people" dimension through the deployment of OCM frameworks, tools, and techniques under the overall governance umbrella of program management. In *Managing Change in Organizations: A Practice Guide,* PMI (2013) enumerates change management as an essential organizational capability that cascades across and throughout portfolio, program, and project management. The PMI guide stresses that change in organizations is delivered through program and projects, and successful organizations lead change by managing their programs and projects effectively. The essence is in ensuring that the OCM plan is integrated into the transformation program plan.

The project plans for all six of the dimensions of program management (i.e., strategy, people, process, technology, structure, measurement) to be incorporated into the integrated transformation program plan. A holistic master OCM plan covers and integrates the multiple OCM sub-areas. Examples of sub-areas include change strategy, change impact, communication, stakeholder engagement, readiness, training, etc., and typically there will be project plans for each of these areas, as these are large areas needing to be planned and executed.

Objective

The strategic objective of a program launched to transform a business is about reinventing the business, which requires the organization to change. The business outcomes and solution (perhaps a new system, product, service, or process) resulting from the business transformation program have to be adopted by the end user community, which could be internal and external. Some of the business changes brought forth by the program will typically encounter resistance from different pockets within and outside of the enterprise. A structured approach is needed to overcome the resistance and for the organization to embrace the delivered solution.

A business transformation program needs a robust OCM plan to succeed, and such a plan is a subset of the integrated transformation program plan detailed later in this chapter. OCM details the process for anticipating change barriers and planning interventions to facilitate the ongoing forward movement of the transformation program. The strategic objective of OCM is to make the program a success by effectively driving the change and facilitating change adoption. The program management team will apply OCM methods and techniques to consummate the strategic objective by architecting and executing an end-to-end OCM plan, which does the following:

- Defines change and formulates change strategy
- Determines the impact of change to business and secures involvement of stakeholders
- Increases stakeholder awareness and readiness by engaging, communicating, and training
- Evaluates the organizational structure and roles and redesigns them if needed
- Designs and implements transition management strategy and transition plan

Approach

A proven OCM framework to effectively define and manage the significant business change brought about by a transformation program is depicted in Figure 7.2. The three iterative phases driving the end-to-end organization change management process are:

Phase 1: Envision. This phase is the big picture of vision and strategy.
Phase 2: Plan. This phase spells out the course of action.
Phase 3: Execute. This phase is the action taken to realize and sustain the strategic objectives.

- Phase 1: Envision
  The OCM strategy for the business transformation program is formulated in this envision phase. The completion of the first phase of the program management life cycle delivers the program strategy which serves as the input to the OCM strategy. The program management team has to ensure the OCM strategy is aligned to the program strategy. During the envision phase, the strategy for each of the eight OCM sub-areas is crafted and an overall cohesive strategy is put in place.
- Phase 2: Plan
  During this second phase, an organization change management plan is developed by tapping into the organization change capability expertise within the transformation program management team.

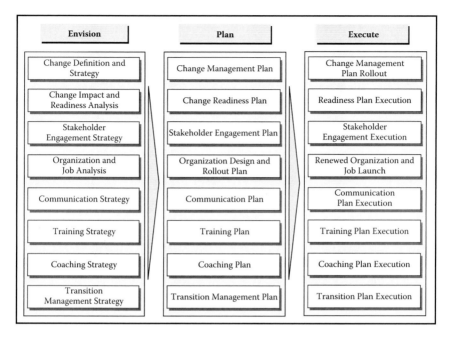

Envision	Plan	Execute
Change Definition and Strategy	Change Management Plan	Change Management Plan Rollout
Change Impact and Readiness Analysis	Change Readiness Plan	Readiness Plan Execution
Stakeholder Engagement Strategy	Stakeholder Engagement Plan	Stakeholder Engagement Execution
Organization and Job Analysis	Organization Design and Rollout Plan	Renewed Organization and Job Launch
Communication Strategy	Communication Plan	Communication Plan Execution
Training Strategy	Training Plan	Training Plan Execution
Coaching Strategy	Coaching Plan	Coaching Plan Execution
Transition Management Strategy	Transition Management Plan	Transition Plan Execution

FIGURE 7.2

Program management: enables integration with organization change management to effectively drive business change.

The output of the envision phase forms the basis for the plan phase. The OCM plan covers all the OCM sub-areas and this plan is a subset of the comprehensive program management plan for the transformation program.

Phase 3: Execute

The execution phase is the implementation of the end-to-end OCM plan. The monitoring of execution of the OCM plan will drive any changes and enhancements needed in the OCM plan. The resource needs for the execution phase is much higher than those for envision and planning phases. The timely completion of the activities composing the OCM plan facilitates the realization of the strategic objectives of the business transformation.

The eight OCM sub-areas (or building blocks) that cycle through these three phases are:

1. Change definition, strategy, and plan
2. Change impact and readiness

3. Stakeholder engagement
4. Organization analysis
5. Communications
6. Training
7. Coaching
8. Transition management

Next, the sub-areas are covered first at a high level and then followed by a sampling of additional frameworks that specifically play a key role in stakeholder engagement.

1. **Change definition, strategy, and plan:** Program management team collaborates with subject matter experts in creating a picture of "what's changing," which is the change definition. The team then articulates the game plan of "how to" change and creates a sense of urgency in formulating the change strategy. The change management plan that dovetails into the business transformation program's plan is developed and implemented.

2. **Change impact and readiness:** This OCM sub-area describes the impact of the business transformation program from a people, process, and technology perspective and who (internal plus external) is getting impacted. The nature, extent, and timing of impact at an organizational and individual level are analyzed. Change readiness assessments and plans are constructed and implemented.

3. **Stakeholder engagement:** In the stakeholder engagement building block, the stakeholders are identified and analyzed. A stakeholder engagement strategy and a stakeholder engagement plan are devised. The engagement strategy is implemented. Getting involvement and engagement of program stakeholders is imperative for program success. Stakeholder engagement is further detailed later in this chapter.

4. **Organization analysis:** An assessment of the organizational policies, roles, reporting hierarchy, and job descriptions is completed as part of the organization analysis component of OCM. The assessment drives adjustments to the current organization model. The renewed organization is smoothly and effectively launched by executing the roll-out plan devised for the same.

5. **Communications:** Based on the communication requirements of stakeholders, a communication strategy and a communication plan are devised. The communications plan is rolled out and adjusted as

needed for the intended communications objectives to be fulfilled. Proven communication frameworks to drive effective end-to-end communications are discussed later in this chapter.

6. **Training:** Based on the organization change readiness assessments and identified training needs, a training strategy and a supporting training plan are developed. A training plan is executed. This OCM building block plays a significant role in increasing and sustaining the acceptance of the business transformation program. A comprehensive framework on training is covered later in this chapter.

7. **Coaching:** The coaching strategy is designed to take care of the one-on-one coaching needs of the impacted stakeholders. A coaching plan to implement the coaching strategy is created. The program management team coordinates and oversees the execution of the coaching plan.

8. **Transition management:** For the effective transition of the business transformation program's output and sustainment of the program outcomes, a transition management strategy is formulated. The transition management layer within the organizational change management framework focuses on the "people" base or dimension. The "transition to operations and close program" process of the program management life cycle is designed to cover all the bases to ensure a smooth transition. The transition management plan is created and implemented to smoothly transition the program to the operational functions within (and outside, if applicable) the enterprise from a people perspective.

Drivers

The transformed business will position employees to ultimately deliver higher value in rapid time frames to its customers. The attainment of future state upon completion of the transformation program will help in improving the overall productivity of the organization. For enterprise level business transformation initiatives, it is imperative for the organizational and individual goals to be in alignment to achieve sustainable change. People tend to resist change, regardless of the nature and the reason for the change. Resistance might be linked to personal factors, fear of the unknown, organizational issues, perceived loss of job security, and lack of communication and training.

Organization change management is a process to get the organization and individual personnel ready for handling the business change. It facilitates getting the commitment and buy-in at all levels to the changes driven by the transformation program. Application of the OCM techniques can change the mindset of certain personnel, track the progress of all changes, solicit feedback, and trigger timely corrective actions. OCM calls for engaging employees, customers, and business partners through the course of the business transformation journey. Stakeholders become part of the future state solution and embrace the behavioral changes needed for realization and sustainment of the planned business outcomes, benefits, and results. In the PMI sponsored white paper titled "Organizational Capacity for Change," Harrington and Voehl (2014) emphasize that an organization can ensure program success by building its capacity for change, while ignorance of talent management, standardized project management, and strategic alignment can constrain an organization's change capacity.

ORGANIZATION CHANGE READINESS

Given the dynamic business climate and numerous external forces, organizations have to always be on the lookout for the strategic changes they need to make to have a thriving business. The general change readiness ability of the enterprise, business unit, or function can be known to a certain extent by looking at historical data on how the enterprise has adapted in the past. In the context of an in-flight business transformation program, one method of getting a pulse of the change readiness capability is through the development and administration of tailored change readiness survey instruments to the stakeholders in the impacted areas.

Analysis of the findings of such surveys will identify key organizational and individual level enablers and barriers to change. The risks to the success of the transformation program can be identified, and mitigating strategies can be developed. Once the current state of organization change readiness for the business transformation program is understood, the change readiness model (Figure 7.3) can be utilized to close out the gaps. Change readiness modeling plays a crucial role in creating a culture of acceptance, improving the overall readiness, and setting the stage for the program to realize and sustain the defined business outcomes. Change readiness modeling entails:

- Formulating the change readiness strategy
- Creating the change readiness plan

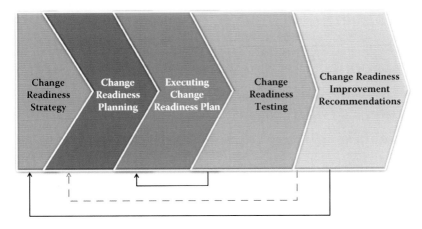

FIGURE 7.3
Organization change readiness model.

- Executing the change readiness plan
- Testing change readiness by working through the most likely future state business scenarios
- Modifying change readiness by incorporating feedback from testing

TRANSFORMATION PROGRAM PLANNING TECHNIQUE

Overview

Transformation program planning is a set of techniques for finalizing the program approach and detailing the work to be performed to complete a program. Program architecture provided the high level approach for the business transformation program, and program planning takes that further in building out the detailed approach in the program plan. In addition to program architecture, the other major inputs to program planning are program strategy, program road map, and program charter. The program plan should map out the work pertaining to the dimensions of strategy, people, process, technology, structure, and measurement.

As an illustration, if a business transformation program entails radically redesigning major business processes and automating these processes, the activities detailed in the program plan need to reflect the same. The transformation program plan captures work and deliverables associated with the program management life cycle phases. Finally, the program plan reflects the key program management processes, and examples of these processes include governance management, integration management, scope management, time management, financial management, risk management, and business outcome management.

Objective

The primary goal is to develop an end-to-end program plan with the inputs of project managers and subject matter experts, secure executive sponsor sign-off on the program plan, and baseline the program plan. The program plan guides the execution of the transformation program and helps the program team succeed. The level of detail in the program plan can vary depending on the nature, complexity, size, structure, and duration. Desired capabilities in the program architecture are typically broken into functionalities in the program charter, which is often referred to as the program description document. The business and technical requirements from the elicitation processes are usually captured in the business requirements document deliverable. On technology projects, the functional system design artifact is developed based on the signed off business requirements document. The program plan needs to clearly specify the activities, owners, timing, and dependencies for key deliverables.

Approach

Program planning in general, and specifically building an end-to-end program plan for a business transformation program, is not just a matter of laying out a set of cookie cutter processes and activities. The transformation program planning process comprises the following steps:

- Understand the problem
- Understand the scope
- Create the program plan
- Finalize the program plan
- Baseline the program plan
- Communicate the signed-off program plan

Step 1. Understand the Problem

A thorough and accurate understanding of the problem statement is necessary to develop a transformation program plan. The problem can be validated by performing the following:

- *Evaluate performance needs*: A business transformation program is usually necessary because a performance gap has been identified where the current level of performance falls short of what is necessary or desired. The enterprise must change itself in some way to raise its level of performance. This gap may be shaped by the enterprise's external environment. The performance need should focus on a set of desired business outcomes. These can take the form of a target that may be broad or narrow. An example of a broad business outcome target is to continue to be the leading provider of customer service within the industry. An example of a narrow business outcome target is to improve performance to justify keeping the order fulfillment process in house.
- *Understand initial constraints*: Another set of factors to consider are initial limitations or constraints. Factors such as the projected cost of a transformation program, the urgent need for results, or impending regulatory or structural changes to an industry may have a dramatic impact on the constitution and ultimate success of a transformation program. Identifying and addressing constraints at the beginning of the transformation program planning process can save time, frustration, false starts, and wasted work. If the program has insufficient funding, little or no sponsorship, or many powerful enemies, it may not be possible for it to succeed. One should address these issues as objectively as possible, because a failed business transformation program will waste precious time, money, and resources. It can also give transformational change efforts a bad name within the enterprise and make it even more difficult for business transformation projects to succeed in the future.

Step 2. Understand the Scope

Business transformation programs can be as broad and far reaching as completely redefining the kind of business the enterprise is in, or it can focus narrowly on a particular business process. The scope identifies the breadth and depth of business change the program is expected to accomplish. The different kinds of transformation programs are explained as follows:

- *Enterprise transformation program*: This program involves the innovation of the entire enterprise, including its mission and values. Such a strategic program will completely transform the entire enterprise, sometimes moving it into a different industry altogether.
- *Business unit transformation program*: This program entails the reinvention of a designated business unit (or line of business or business division) within the enterprise. Such a program is typically begun in response to a dramatically weakening performance of the business unit, and there is a general recognition and agreement that only the most radical redesign of the unit will save it.
- *Business function transformation program*: This program type has a more narrow focus, with the strategic objective being radical redesign of a business area or function, effectively starting with a blank piece of paper. Such programs are the most common kind and result from examining a business process contained within a functional area of an enterprise, or it can entail the innovation of a cross functional process, e.g., order to delivery, customer service, product development, brand management, claims processing. While such programs can be disruptive, typically the overall enterprise is not affected. Program risk exists, but because it is limited in scope, risk can be limited.
- *One dimensional transformation program*: Such a program brings about changes that can be implemented fairly quickly and inexpensively, with relatively little effect on the rest of the enterprise. For example, a business process improvement program might be designed to transform significant business processes. Another example of a single dimensional program is an organization change management program launched to train a vast end-user population on a new enterprise system that is being implemented. Though such one dimensional programs give good results with relatively low cost, it is often a Band Aid patch to a larger problem. Unfortunately, such a one dimensional program is often insufficient to gain a lasting competitive advantage.

Step 3. Create the Program Plan

As mentioned previously, at a high level, the transformation program plan brings together the six program management dimensions, program management life cycle phases, and program management processes. Based on the nature and complexity of the program, the work associated with the six dimensions could be structured as projects with stand-alone project plans, but they all need to be tied into the end-to-end integrated program plan. Outlining in detail the work that needs to be done is a difficult and time consuming task, but it is important in understanding the charter and scope of the program. Every transformation program has to deal with major cross functional issues, risks, changes, resource limitations, and constraints. A good program plan has several components, with the key ones being tasks, schedule, resources, deliverables, dependencies, and milestones.

- *Key inputs and project plan roll-ups*: The following completed and signed-off deliverables are the key inputs to the program plan development exercise: program architecture, program strategy, program road map, and program charter or program description document. Though plan development is an iterative exercise, the final program plan is based on the baselined program charter. A business transformation program is typically structured to have multiple projects beneath, and at times there could be projects and programs beneath. Figure 5.2 shared an example of the typical program organizational structure, and Figure 4.1 highlighted the sponsorship structure for such a program. Some of the sections of the integrated program plan constitute higher level roll-ups of the project plans that will be driving the projects comprising the transformation program. The project plans should have been finalized, signed off, and baselined prior to them getting rolled up for inclusion in the program plan.
- *Program management processes*: The program manager and program management team owns the fourteen program management processes highlighted in Figure 7.4. The eight processes constituting the program management life cycle are the highest level processes, and these are the verticals. The fourteen horizontal processes running through the four phase program management life cycle can be envisioned as the next level processes. The horizontally depicted

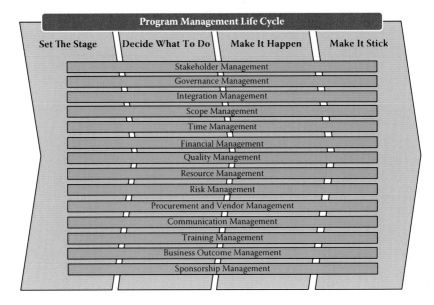

FIGURE 7.4

Program management: realizes vision by implementing strategy.

processes are decomposed into the next level processes, which can be referenced as subprocesses or activities from a terminology standpoint. The synthesis of these horizontally and vertically mapped program management processes provides a comprehensive framework for strategic execution of the business transformation program and implementation of the program strategy. As these processes are discussed throughout this book, they are simply listed here without additional elaboration:

- Stakeholder management
- Governance management
- Integration management
- Scope management
- Time management
- Financial management
- Quality management
- Resource management
- Risk management
- Procurement and vendor management
- Communication management
- Training management

- Business outcome management
- Sponsorship management

Step 4. Finalize the Program Plan

While the approach for creating the transformation program plan has been presented in a fairly linear manner, the actual process is highly iterative. If, after creating the program critical path and building a program schedule, one finds that the program will take too long, one has several choices. Some of the deliverables cannot be created, the level of detail or quality must be lowered, additional resources must be deployed, or the scope revisited. Also, the overall program plan should be reviewed to see if there is any slack in the schedule that can be removed. The answer is not to just throw more resources at the program. The transformation program plan has to be socialized with subject matter experts and key stakeholders. The final program plan on which buy-in has been secured is reviewed with the sponsorship team, and approval is secured.

Step 5. Baseline the Program Plan

The approved program plan is baselined. After baselining, any changes to the program plan will have to go through the standard change control process established for the program. Identical governance applies for the baselined project plans. If there are changes to the baselined project plan—and if those changes impact the baselined program plan—then the program management team has to update the program plan. A standardized process for such iterations between the program management and project management teams working on the program needs to be in place. Controls around the maintenance of the program and project plans also need to be implemented.

Step 6. Communicate the Signed Off Program Plan

The baselined program plan is published and shared with the business transformation program's team and other pertinent stakeholders. Also, the processes for program plan iterations and updates to the program plan are relayed to the right audience. The communication process should have the ability to publish updated program plans right away to those who need it.

				Benefits Realization Planner				
Id	Business Objective	Business Outcome	Benefits Description	Benefits Realization Owner	Benefits Realization Target Date	Key Value Indicators	Value Metrics	Measurement Method
1								
2								
3								
4								
5								
6								

FIGURE 7.5
Program benefits realization planner.

BENEFITS REALIZATION PLANNING

The business outcome management process drives the development, execution, and monitoring of the benefits realization plan. The benefits realization plan is developed as part of the transformation program plan creation and finalization process. The benefits realization plan is a critical deliverable, as it ensures there is stakeholder alignment on the business benefits the business transformation program is expected to realize. It also indicates the time frame for realization of the expected benefits. The benefits realization plan needs to be maintained as the program progresses through the life cycle. The program manager ensures the existence of ownership, process, governance, and communications on the updates to the program's benefits realization planner. Figure 7.5 is a self explanatory sample of a planner many organizations have successfully used on business transformation program.

Case Study: Business Transformation Program to Launch a New Product

CONTEXT

A large and highly reputed U.S. national health care enterprise with an integrated health care delivery model wanted to catch up with competitors who had a more diversified health plan product portfolio. With continuing major overhaul of the U.S. health care system, the senior management at this health care enterprise saw the tremendous opportunities that existed in the marketplace for it to exploit the integrated business model, which is a strategic differentiator.

BUSINESS PROBLEM OR OPPORTUNITY

The gap in the health plan product portfolio in relation to the competition was resulting in this well established enterprise losing out on some strategic opportunities. To close this gap, the organization launched an enterprise level transformation initiative to design, pilot, and roll-out a new product nationally. The complexity of integrating the new product offering within the existing operating infrastructure was high. The health care enterprise started experiencing a slippage in the new product roll-out timeline, as this transformational change necessitated readiness in multiple functional areas in multiple geographic regions. The enterprise was challenged with stand alone silo based work happening in various areas and the lack of a coordinated product launch effort.

SOLUTION

The program team dedicated to this transformation program validated that the conceived product was in line with the market needs and had a nationwide appeal and demand. The program management team structure had a national and regional layer. The program manager with the national responsibility engaged with the regional teams to mastermind the program architecture, which provided the platform to drive alignment and integration between the national and regional work. A program road map was developed sequencing the work to be completed by the products, process, systems, organization readiness, and benefits realization measurement teams. The national program management team prepared an integrated, end-to-end transformation program plan that captured the program critical path and the work to be completed to enable a successful launch of the new health plan product. A 90-day turnaround plan as well as a benefits realization plan was also put in place in devising the program plan.

BUSINESS OUTCOMES AND BENEFITS

The enterprise realized the envisioned future state of the new product being available on a national basis per the original timeline. The new product launched successfully as a trained workforce to sell, enroll, and administer the product was established prior to the official launch. The team bid on and clinched the opportunity with a Fortune 50 bank to offer the health plan to the employees of that bank. The investment made by the premier health care enterprise in planning and creating an integrated approach to take a new product to market yielded a very good return.

SUMMARY

Based on the key transformation program planning concepts and covered processes, one can see that program planning requires significant effort and acumen. The five step program architecture process establishes the bridge between business strategy and implementation of that strategy. The program vision aligns the program to the strategic direction of the enterprise. The output of the program architecture is the key input to the ten program management life cycle. The program strategy and program road map drive the transformation program plan to ensure continued alignment. The bottom up approach of build out of the project plans and higher level aggregation of those detailed project plans into the program plan facilitates the buy-in by the core program team, which has the responsibility throughout the program management life cycle.

The first two phases of program management life cycle are essentially planning phases and the culmination of those results in fruition of a cohesive, integrated program plan. The transformation program planning technique assists in the development of a comprehensive program plan, which has numerous component plans including the business outcome realization plan. Fusion of the organization change management (OCM) plan into the overall program plan lays the platform for garnering engagement and commitment of the stakeholders to the program agenda. The lack of integration and alignment of OCM work with the rest of the work driven by the program management life cycle is a recipe for failure of the business transformation program. An end-to-end holistic and integrated approach will result in attainment of the future state, accomplishment of the strategic objectives and sustainment of the defined business outcomes.

REFERENCES

PMI. 2013. *Managing change in organizations: A practice guide.* Newtown Square, PA: Project Management Institute.

Harrington, H. James, and Frank Voehl. 2014. *Increasing change capacity and avoiding change overload.* PMI white paper. Organizational capacity for change. Newtown Square, PA: Project Management Institute. http://www.pmi.org/~/media/PDF/Publications/Org_Capacity_for_Change_wp_final.ashx

8

Drive Strong Partnership and Stakeholder Engagement

Large scale, complex programs to transform an enterprise, business unit, or business area are most successful when the program management team proactively collaborates with the program stakeholders through the program management life cycle. Trusted relationships have to be cultivated and close partnerships developed as program managers drive the execution of the transformation program to move the business from its current state to the envisioned future state. At the end of the day, a key success metric for the program is the adoption of the attained future state by the program stakeholders. The program manager leading the business transformation effort facilitates embracement of the future state through effective ongoing stakeholder engagement.

The discipline of organization change management was introduced in Chapter 7. Stakeholder engagement is one of the eight sub-areas encompassing change management. A strategy, road map, plan, and process for stakeholder engagement—coupled with the flexibility needed to engage with internal and external stakeholders at multiple organizational levels and varying personalities—is necessary. As a typical transformation program is driving significant business change and impacting many stakeholders, the management of individual level and organizational level expectations of stakeholders through effective ongoing engagement with them is a science as well as an art.

The following topics are described in this chapter with the help of supportive illustrations, including a real world case study:

- Stakeholder expectation management technique: Overview, objective, and approach
- Stakeholder assessment

- Stakeholder engagement strategy
- Stakeholder engagement: Planning, executing, and monitoring
- Stakeholder engagement through communications vehicle
- Stakeholder engagement through training vehicle
- Stakeholder engagement through coaching vehicle
- Business outcome delivery enabler
- Case study: Transformation of policy servicing at enterprise level

STAKEHOLDER EXPECTATION MANAGEMENT TECHNIQUE

Overview

The process of developing trusted relationships and partnerships with stakeholders comprises a series of effort intensive steps. In order to succeed in effectively managing stakeholder expectations from the start to finish of a transformation program, the expectations need to be well understood by the program management team, and those have to be validated on an ongoing basis. A blueprint for identifying and managing stakeholder expectations is diagrammed in Figure 8.1. The stakeholder expectation management blueprint has the following four elements:

- Change readiness model
- Communication model
- Training model
- Coaching model

The change readiness model was covered in Chapter 7. The communication, training, and coaching models are covered later in this chapter. The stakeholder expectations on a complex business transformation program can be effectively managed by these four elements iterating through the following stages:

- Strategy
- Planning
- Executing
- Monitoring
- Improving

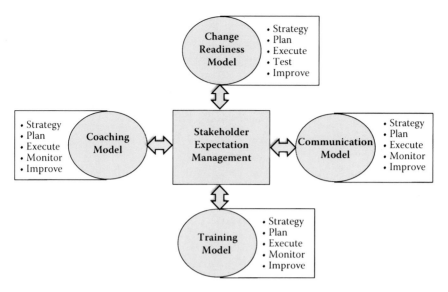

FIGURE 8.1
Stakeholder expectation management model.

Objective

The ultimate goal of the stakeholder expectation management process is to make the stakeholders "own" the transformation program, which would drive their behavior to embrace and adopt the future state created upon completion of the program. The expectations of stakeholders need to be closely managed through the program management life cycle to align with what the program will deliver. Targeted objectives are:

- Develop a stakeholder engagement strategy and implement stake-holder engagement strategy
- Create and implement an organization readiness strategy
- Architect and execute communication strategy
- Build out a training strategy and implement training strategy
- Formulate and roll-out a coaching strategy

Approach

Stakeholder expectation management involves the following:

- Stakeholder assessment
- Stakeholder engagement strategy

- Stakeholder engagement planning, execution, and monitoring
- Stakeholder engagement through communications vehicle
- Stakeholder engagement through training vehicle
- Stakeholder engagement through coaching vehicle
- Stakeholder engagement through communications vehicle
- Stakeholder engagement through training vehicle
- Stakeholder engagement through coaching vehicle

STAKEHOLDER ASSESSMENT

Stakeholders are individuals and groups, both internal and external, that impact or get impacted by the business transformation program and have a stake in the success and behavior of the program. Stakeholders include (among others) program team members, customers, employees, suppliers, regulators, investors, the community, and business partners. The transformation program charter deliverable generated by the "define program charter" process of the program management life cycle is a good starting point for identifying the stakeholders. As the program management team furthers the program management life cycle, additional stakeholders can be identified. An analysis of stakeholders is done to qualify and quantify the impact and the level of influence they can have over the business change the transformation program is originating (Figure 8.2).

Stakeholder Identification and Analysis								
Identify Stakeholders			Analyze Stakeholders					
Id	Name	Role	Nature of Impact	Level of Impact	Understanding of Impact	Level of Readiness	Level of Support	Level of Influence
1								
2								
3								
4								
5								

FIGURE 8.2
Stakeholder assessment.

STAKEHOLDER ENGAGEMENT STRATEGY

The stakeholder assessment output and additional insight obtained in going through the stakeholder assessment exercise enables the development of an effective strategy to engage with the transformation program stakeholders. Given that one could have a large number of stakeholders of different types, the premise for effective engagement is to categorize them into logical groupings and have a tailored plan for engaging with stakeholders in each grouping. The stakeholder categorization is accomplished by mapping two dimensions, which results in the grid depicted in Figure 8.3.

In Figure 8.3, the horizontal axis of the stakeholder map is the "level of impact" on the stakeholder by the program, and the vertical axis is the "level of influence" the stakeholder can have on the program outcomes. The four quadrants denote the four logical groupings, and each grouping will have its tailored engagement strategy and supporting plan to facilitate an overall optimal engagement of all the program stakeholders. The stakeholder engagement strategies are:

- Make acquaintance
- Form alliance
- Cultivate relationship
- Build trust

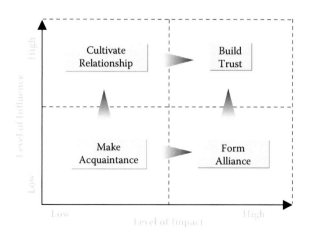

FIGURE 8.3
Stakeholder engagement strategy.

STAKEHOLDER ENGAGEMENT: PLANNING, EXECUTING, AND MONITORING

The program management team utilizes the outputs of stakeholder assessment and stakeholder engagement strategy to devise the game plan to effectively engage with the stakeholders through the business transformation program. Stakeholder engagement objectives and desired engagement outcomes for each engagement strategy are determined. The stakeholder engagement plan has to be comprehensive to ensure that no stakeholders have been missed. It not only has to map out the key details (i.e., objective, outcome timing) for each engagement, but also the engagement vehicle and logistic preferences to facilitate optimal engagement.

Socializing the engagement plan with the stakeholders and securing their buy-in positions the program team to successfully execute the plan. As the stakeholder engagements happen in line with the plan, the realized engagement outcomes should be noted and compared to the original planned outcomes. From a continuous improvement perspective, such monitoring and tracking of execution of the engagement plan assists with refining the stakeholder engagement plan. Figure 8.4 highlights a sample template for stakeholder engagement planning, execution, and monitoring.

Stakeholder Engagement - Planning, Executing, and Monitoring								
Plan Stakeholder Engagement					Execute and Monitor Stakeholder Engagement			
Id	Stakeholder Name	Stakeholder Role	Engagement Objective	Engagement Strategy	Desired Outcome	Engagement Timing	Engagement Outcome	Realized vs Desired Outcome	Alternate Approach
1									
2									
3									
4									
5									

FIGURE 8.4
Stakeholder engagement.

STAKEHOLDER ENGAGEMENT THROUGH COMMUNICATIONS VEHICLE

Stakeholder expectations can be effectively managed through timely, proactive stakeholder engagement. The stakeholder engagement plan formalizes the approach to managing stakeholder expectations and places a rigor in the engagement process. The actual engagement with the stakeholders happens through multiple vehicles, with communications being one of the most important. Examples of stakeholder engagement vehicles are communications, training, readiness assessments, and governance forums. Figure 8.5 spotlights the communications processes of:

- Devising the communications strategy
- Creating the communications plan
- Executing the communications plan
- Monitoring the communications plan
- Conducting periodic surveys to get feedback on communication
- Refining communications strategy, plan, and execution based on feedback

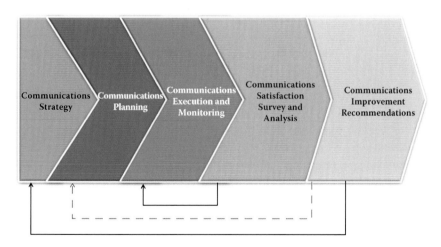

FIGURE 8.5
Stakeholder engagement: communication life cycle.

The communication life cycle is the end-to-end communications framework supporting the business transformation program. It describes how and when the transformation program vision, objectives, implementation status, and rationale for critical decisions will be communicated throughout the organization. Using information assimilated within the communications plan, targeted messages will be delivered on a regular basis through effective communication channels. The communications life cycle ensures that key stakeholders will be engaged for maximum leverage and favorable impact of the transformation program.

A communications vehicle increases the awareness of the transformation program, secures buy-in and involvement, minimizes the resistance to change, and ultimately positions the organization for successful roll-out of the business transformation. PMI's *Pulse of the Profession In-Depth Report: The Essential Role of Communications* (2013) analyzed the data collected from 742 project management practitioners to conclude that more effective communications leads to improved project and program management, more successful projects, high performance, and fewer dollars at risk.

STAKEHOLDER ENGAGEMENT THROUGH TRAINING VEHICLE

The training life cycle is the end-to-end training management framework supporting the business transformation program. Stakeholder engagement though the training vehicle educates stakeholders about the business transformation program, explains the changes the program will bring about, and prepares them to thrive in the post transformed future state. Training empowers stakeholders, develops their necessary skills, and facilitates their usage and adoption of the solution (e.g., a new product, service, process, or system) delivered by the business transformation. The training life cycle (Figure 8.6) encompasses the following:

- Develop the training strategy
- Build out the training plan
- Implement and monitor the training plan
- Perform periodic surveys to proactively solicit feedback on training
- Adjust training strategy, plan, and execution based on feedback

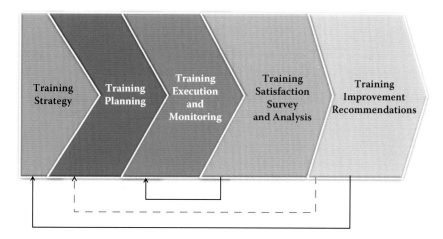

FIGURE 8.6
Stakeholder engagement: training life cycle.

STAKEHOLDER ENGAGEMENT THROUGH COACHING VEHICLE

The two earlier vehicles of communication and training for effectively engaging with and managing stakeholder expectations can be applied quickly towards a large set of program stakeholders. The coaching vehicle is typically applied for effectively engaging with eleven select stakeholders of the program on a one-on-one basis. As a result, the coaching vehicle is a good complement to the communications and training vehicles. Often, coaching is leveraged to manage the expectations that executive level stakeholders have of the business transformation program. Coaching can address the unique needs of such stakeholders with a high level of influence and/or high level of impact on the program. In addition, executive level stakeholders have limited time, high decision making authority, and their behaviors and actions are observed by many other program stakeholders. A comprehensive coaching program can be designed and implemented to garner the sponsorship, partnership, and support of such high influential key stakeholders.

The strategy and plan for stakeholder expectation management through the coaching vehicle can be tailored to address the specific needs of the targeted individual. The stakeholder audience for engagement though coaching need not be just senior level executives. The grid in Figure 8.3 highlighted the four groupings of stakeholders. A tailored coaching program can be architected and deployed for select individuals in each of these groupings. The development, implementation, and refinement of a coaching program entail the following:

- Formulating the coaching strategy
- Developing the coaching plan
- Executing the coaching plan
- Assessing the coaching plan
- Improving the coaching plan

BUSINESS OUTCOME DELIVERY ENABLER

Some of the deliverables from management of a transformation program can play a significant role in facilitating the achievement of defined business objectives. As the program manager(s) work closely with the core program team in defining and finalizing the deliverables stack for the program, the due diligence exercise around the "value" of each deliverable is an invaluable one. The nonperformance of such an exercise can result in a nonoptimal utilization of the program resources. The program team has to be focused on producing deliverables that the customers of the program perceive as a value addition. Also, the basis for value determination is how that deliverable dovetails with the business outcomes the stakeholders are expecting from the transformation program.

In assessing which deliverables carry the highest value in the eyes of the stakeholders, the following factors come into play:

- Targeted audience for the deliverables
- Process followed in developing the deliverables
- Stakeholder expectations of that deliverable
- Content as well as the level of granularity in the deliverable
- Deliverable timing
- How some of the deliverables relate to each other
- How well the deliverables have been socialized with the key stakeholders

Sample Deliverable	How Deliverable Facilitates Outcome Delivery
Program Strategy	• Clear understanding of vision by business and technology stakeholders • Operational and technology driven strategy that provides an integrated view across business and technology
High Level Business Process Models	• Identifies opportunities for streamlining processes • Builds a consensus view across business and technology of the future state business processes
High Level Technology Architectures	• Establishes clear direction of the future state for IT to move toward • Aligns application, data, and information architectures to business drivers
High Level Business Case	• Brings out the business justification for large implementation projects • Provides information needed to request funding for projects
Program Road Map	• Actionable implementation plan that is prioritized from a business angle • Identification of quick wins

FIGURE 8.7
Program deliverables should enable outcome delivery.

Figure 8.7 illustrates how to describe to the stakeholders that the program management deliverables enable business outcome delivery.

Case Study: Transformation of Policy Servicing Function at Enterprise Level

CONTEXT

One of the top three providers of mortgage insurance and credit enhancement products to the global market was struggling with a highly inefficient, error prone, and cost intensive policy servicing platform. This highly successful public company had not made enough investments over the years in upgrading the technology backbone that supported the policy servicing function, and the severity of pain points continued to mount as the company's product portfolio expanded.

BUSINESS PROBLEM OR OPPORTUNITY

The mortgage insurance company had an obsolete, nonintegrated technology environment that was not only inefficient, but also a major bottleneck to accomplishing the company's business goal of worldwide expansion. The company's policy servicing processes were not standardized across the product lines that served the domestic and international markets. The lack of standardized processes coupled with some manual processes was becoming a significant operational drain. As the end users in the company's policy servicing organization and information systems function were experienced and comfortable with the current state, there was organizational resistance to overhauling the servicing business processes and systems.

SOLUTION

The company assembled a small team of technology experts and functional subject matter experts to evaluate the current state. The technical team designed the new technical architecture that accommodated the desired technical capabilities. The business team devised the business case for transforming the policy servicing function by using the program value justification technique covered in Chapter 4. The program team leveraged these "quick wins" and secured sponsorship from a steering committee comprising functional heads. The development and effective communication of a "change campaign" to impacted stakeholders and usage of other organization change management (OCM) tools enabled the program to gain visibility, support, and momentum.

The policy servicing transformation program team designed a stakeholder engagement strategy and plan. A blueprint of the future state of policy servicing, a high-level program road map, and a detailed transformation program plan to attain the future state were created in engagement with key stakeholders. The embedded OCM plan in the transformation program plan captured the eight OCM sub-areas showcased in Chapter 7. As the program execution progressed, the program team leveraged the stakeholder expectation management model to gain acceptance for the business process and technological changes underpinning the future state of the mortgage insurance company's policy servicing function.

BENEFITS OR BUSINESS OUTCOMES

The desired future state of the policy servicing was realized in eighteen months, and the stakeholders embraced the implemented solution. The end users enjoyed the higher automation, and the technical team had better system support tools. Most of the historical constraints the products group encountered in developing mortgage insurance products for the United States and international markets were eliminated. The mortgage insurance company could now compete more effectively in the faster growing international market. The concerted effort of the program management team in driving partnership amongst the company's different subgroups and engaging effectively with stakeholders throughout the program management life cycle resulted in the successful transformation of the policy servicing function.

SUMMARY

The hallmark of many successful business transformation programs is the active engagement and involvement of stakeholders throughout the transformation life cycle. Early on, the program team has to perform a comprehensive and diligent stakeholder identification and assessment exercise. A tailored stakeholder engagement strategy and stakeholder engagement plan has to be architected. The diligent implementation of the stakeholder engagement plan will aid the transformation program team in getting stakeholder participation and developing trusted partnerships.

Communication, change readiness, training, and coaching not only enable stakeholder expectation management, but they also support the change process in transitioning from the current state to the future state. The lack of inadequate participation and buy-in by the key stakeholders is a major red flag that needs to be overcome at the earliest. The periodic monitoring of effectiveness of stakeholder engagement is also crucial as it is not uncommon for the stakeholder landscape to be changing. For the transformation program to successfully deliver and sustain the change, stakeholder engagement at all levels is necessary with the executive sponsor supporting and championing and supporting the change.

REFERENCE

PMI. 2013. *The essential role of communications.* Pulse of the Profession In-Depth Report. Newtown Square, PA: Project Management Institute.

9

Provide Leadership Across All Levels

The nature and level of leadership on a program designed and launched to transform the entire business or significant parts of the business are primary determinants of the program being successful by achieving the intended business goals. Given organizational matrices, work complexities, and corporate politics, the program management team personnel rely heavily on their leadership traits to deftly guide issues and manage risks through multiple organizational levels and across functions. Leaders can model the behavioral changes needed to embrace the changes the program is driving.

Leaders should effectively articulate the benefits of the transformation program to stakeholders and get them on board. By their very nature, transformation programs are complex and aim to make significant changes to the business. Without a multidimensional, multilevel leadership and cross functional leadership, the program will inevitably be a total failure or will not accomplish all the strategic business objectives. Program management plus program leadership are critical success factors for facilitating work integration, stakeholder engagement, objective alignment, organizational change readiness, and benefits realization.

The following topics are described in this chapter with the help of supportive illustrations, including a real world case study:

- Multilevel program leadership model
- Multidimensional program leadership model
- Program leadership and risk mitigation
- Transformation program leadership: Planning and delivery stages
- Leadership of program management processes
- Program leadership versus program management
- Balancing program leadership and program management

- Program leadership through office of business transformation
- Drivers for the office of business transformation
- Significance of leadership in program communications
- Benefits realization leadership
- Case study: Business transformation initiative on privacy and compliance

The lack of leadership structure on a transformation program or one with voids in program leadership skills jeopardizes the mission of that program. For a complex program to be successful in realizing the program vision, achieving business objectives, and delivering the targeted business outcomes, the need for proactive leadership through the course of the program is critical. As the program team will confront cross divisional and cross functional barriers and challenges, the program management team has to determine the leadership needs and put in place multiple leaders in the early stages of the transformation program. Program leadership is needed at multiple levels, in numerous areas, of different styles, and in varying levels of involvement to successfully chart the course from current to future state, which results in major business change. Leadership in stakeholder engagement and securing stakeholder buy-in will ensure that the expected program outcomes are realized at each phase of the transformation program.

MULTILEVEL PROGRAM LEADERSHIP MODEL

The program organization model lends the structure and governance needed for a business transformation program to transition an organization from its current to its desired future state. Figure 5.2 in Chapter 5 is an example of a program management model. The designed model needs to be socialized and communicated to get the buy-in of stakeholders and raise awareness of the decision making process. The pictorial program organization model depicts the reporting hierarchy and forms the basis for program governance. As it is critical for all program stakeholders to have clarity in roles and responsibilities, these have to be defined, agreed upon, and communicated. For an example of same, please refer to Figure 4.2 in Chapter 4. Multilevel leadership stresses the importance of leadership at each tier of the program organization structure. In other words, leadership on complex

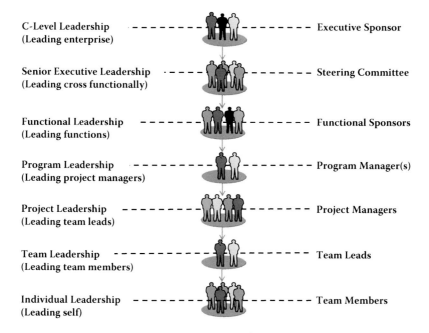

FIGURE 9.1
Multilevel leadership needed for success of transformation program.

programs can't reside only at the highest levels of the program organization. In Chapter 5, the implemented governance and sponsorship models on a large scale business transformation initiative were illustrated. Let's look at the leadership model for the same program (Figure 9.1). This model spells out the multilevel leadership on that successful transformation program. The scope, business context, and level of influence of leadership at each program organization level are different. For example, the program steering committee is responsible to provide leadership over the many functions involved in the business transformation and has the authority to do so. The program managers, on the other hand, are the leadership linchpins in advising, directing, guiding, and supporting the project managers who are leading specific projects with the help of the team leads. The leadership responsibility and accountability at each level is defined, and these align to the authority level for that tier. The design, socialization, and deployment of a multilevel program leadership model will ensure that the leadership needs of the program are fully met for the program management life cycle.

MULTIDIMENSIONAL PROGRAM LEADERSHIP MODEL

Program management of a transformation initiative requires leadership over the six dimensions of the program—strategy, people, process, technology, structure, and measurement. The program charter, scope, approach, and company structure will govern how the integrated program plan is structured to address all of these six dimensions. The multidimensional aspect of the leadership model can be understood by referencing Figure 4.1 from Chapter 4 on sponsorship. In that example, the dimensions of process, technology, structure, and people are reflected in the program delivery tracks, which run across the functions (sales, manufacturing, etc.) in play. In the scenario depicted in Figure 4.1, from a program leadership perspective, leadership is needed over the numerous program delivery tracks as well as functions.

The business change definition and impact analysis work of the program team coupled with the stakeholder analysis provides the key inputs to develop the program leadership model. The model has to line up leaders for each impacted stakeholder group, and it is best to target individuals who have the knowledge, influence, commitment, and authority over that group. All the leaders have to direct, support, embrace, manage, and champion the transformation program for it to fire on all cylinders during the course of the transformation journey. The development and implementation of a multidimensional program leadership model is essential for the transformation program to transition the organization from the current to the desired future state.

PROGRAM LEADERSHIP AND RISK MITIGATION

Dynamic leadership not only enables the realization of desired business outcomes, but it does so without the organization having to undertake huge risks. The organization's capability to deliver the highest outcomes with minimal risks is similar to a financial investment professional striving to deliver the highest returns without betting the farm. Strong leadership identifies and mitigates the key organizational risks that can prevent the attainment and sustainment of the targeted benefits and outcomes.

Program leadership plays a crucial role in managing major program risks, some of which are listed here:

- Lack of prioritization of programs results in too many programs at the same time.
- There is opposition to standardization and a heavy push for tailored processes and systems.
- The platform for transformational change is not anchored against benefits.
- The engagement level of key stakeholders is low.
- Decision making process is slow and/or there is confusion around decision making processes.
- There are more barriers and fewer enablers to the change needed to transform the business.
- There are gaps in organizational capability.
- Simultaneous implementation of major programs leads to resource bottlenecks.

TRANSFORMATION PROGRAM LEADERSHIP

Planning Stage

Top class, consistent leadership on the program is needed for the full life cycle of beginning to end of business transformation. The leadership capability needs and leadership approach vary based on the stage of the program. The focus of leadership during planning is in standing up the program and designing the foundation on which program execution will happen. At the planning stage, the leadership emphasis and direction is on finalization of program architecture, formulation of program strategy, understanding the organizational impact of transformation, and development of the program plan.

During planning, the early work of the core program team is focused on program architecture. As a recap, program architecture builds the bridge between business strategy and strategy implementation. It frames up which of the pool of strategic business objectives is targeted by the transformation program in formation. Chapter 7 expounded on the five processes in program architecture: articulate program vision, assess current state, develop future state, create business case, and design program.

The program management life cycle was introduced in Chapter 1, and the eights processes within the life cycle are touched upon in multiple chapters. The first four processes of the program management life cycle (i.e., formulate program strategy, develop program road map, define program charter, and create program plan) embody the planning stage of the transformation program, with the program architecture work being the input to the "formulate program strategy" process. In Chapter 3, the work carried out under "formulate program strategy" process was explained in reviewing the strategic alignment technique. Chapter 2 described the program charter, and the program plan was covered in Chapter 7. Program leadership during the planning stage has to ensure the creation of a shared program strategy, program road map, program charter and program plan, and the socialization and communication of the same. The shared program strategy articulates at a high level how the program vision will be realized. The charter spells out the problem being solved, what success looks like, and how success will be measured.

The program road map is the high-level implementation plan showcasing the timeline for realization and sustainment of defined outcomes. On business transformation programs, there could be stand alone road maps for people, process, and technology dimensions, with the master road map integrating them to arrive at the high level implementation plan. The program plan is the more granular integrated implementation plan detailing the resources, deliverables, timing, milestones, and critical path. As these planning artifacts will be the basis for and drive the remainder of the program, which typically has a longer duration and a much higher consumption of program approved resources, the leadership during the planning stage of the transformation program is pivotal.

Delivery Stage

The focus of leadership during the delivery stage is on overseeing implementation of the program plan and promptly removing hurdles that prevent the organization from getting to the future state. The last four processes of the program management life cycle (execute program, monitor program delivery, transition to operations and close program, and sustain outcome delivery) represent the delivery stage. During the delivery stage, program leadership is key in getting the organization ready to embrace and accept the change. Leadership in the delivery stage ensures that the alignment to program strategy is maintained as program execution progresses. Similarly, leadership involvement and agreement to any material deviations to the baselined integrated program plan is critical. Transformation program leadership facilitates smooth attainment of the future business state and sustains the desired business outcomes over time.

Leadership is needed not just from a technical business perspective, but also from a human perspective. As a business transformation program is designed to transition the business to a new future state, the impacted stakeholders have to migrate from the old to the new. Resistance to change is normal human behavior, and the resistance can be overcome to a great degree through effective leadership. The benefits of change to the organization and stakeholders can be proactively and periodically communicated by the leaders as part of their championing effort. To increase the level of acceptance at an organizational and individual level, leadership support could manifest in other forms, including sponsorship of training, coaching, recognition arrangements, etc. Regular engagement of leaders with stakeholders and their involvement in the creation of the future business state facilitates a sense of ownership among stakeholders, which is a key for acceptance of the transformation program agendas.

LEADERSHIP OF PROGRAM MANAGEMENT PROCESSES

Matured and institutionalized program management processes and practices are mandatory for the planning and delivery stages to make the transformation a success. In Chapter 7, the fourteen program management processes were identified. These processes are: stakeholder management, governance management, integration management, scope management, time management, financial management, quality management, resource management, risk management, procurement and vendor management, communication management, training management, business outcome management, and sponsorship management.

These fourteen program management processes cut across the program management life cycle and are instrumental in making the transformation program move forward in line with stakeholder expectations. Leadership over the execution of the early activities within each of these fourteen processes ensures that a sound strategy and approach is in place for each of these fourteen processes. A strong, proactive leadership provides the much-needed foundation for the planning stage that was discussed earlier to be successful. Similarly, leadership over the execution of the subsequent activities within each of these fourteen processes positions the delivery stage to perform in line with stakeholder expectations.

PROGRAM LEADERSHIP VERSUS PROGRAM MANAGEMENT

Program leadership and program management are distinct. From an organizational capability perspective, both program leadership and program management are needed to realize and sustain the future state and reap the corresponding rewards and benefits. The leadership and management competencies complement one another. The program management team doesn't need separate individuals to play these roles, and there are overlaps. The difference between program leadership and program management are spelled out in Figure 9.2. Program management and

Leadership	Management
➤ Vision translation	➤ Plan creation and management
➤ Decisions that support vision	➤ Deliverables/milestones tracking
➤ Building high performing team	➤ Enforcing process and standards
➤ Empowering, open culture	➤ Time/cost management
➤ Commitment reinforcement	➤ Boundary management
➤ Motivating and morale building	➤ Issues/risk management
➤ Coaching and mentorship	➤ Managing and reporting status
➤ Managing politics tactfully	➤ Critical path management
➤ Relationship cultivation	➤ Stakeholder management
➤ Expectation management	➤ Resource management
➤ Handling feelings and emotions	➤ Vendor management
➤ Directional alignment	➤ Managing communications
➤ Change champion	➤ Change control management

FIGURE 9.2

Program leadership and program management are complementary.

program team personnel have to constantly and simultaneously play the "two-in-one role" as they move the organization from the current state to the future state. There will be monumental challenges that will have to be overcome from a technical (or business) perspective as well as a people (or culture) perspective. Effective planning and execution of the fourteen program management processes by the program manager addresses the technical side of the business transformation. The program manager has to wear the program leader hat in planning and executing the unstructured work that influences the program's team morale, behaviors, and attitudes. The integrated transformation program plan has to specify leadership related activities (e.g., team building, relationship cultivation), but the implementation of those activities requires the program manager (or program leader) to be a role model for the expected behavior. Transformation program leaders not only communicate the program vision, strategy, values, and benefits, but they also understand and effectively combat the organizational resistance to the changes being driven by the program. As ambiguities, uncertainties, and complexities are to be anticipated on a business transformation program, both program leadership and program management are needed at multiple levels to overcome the resistance to business change at these various levels and build an environment of future state adoption.

BALANCING PROGRAM LEADERSHIP AND PROGRAM MANAGEMENT

The program management team doesn't need separate individuals to play the roles of program manager and program leader, but program management personnel need to strategically balance their efforts in planning and executing management as well as leadership related activities. The program manager must possess the leadership traits to lead a complex, cross functional business transformation program and must have demonstrated application of those traits. The lack of program leadership skills is a strategic risk to the delivery of the program. Figure 9.3 portrays the four program execution capability scenarios that typically play out in the field in the context of a business transformation program.

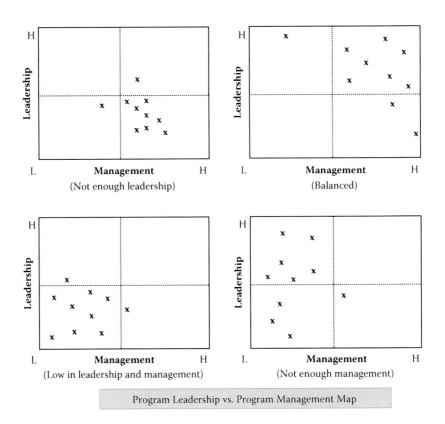

Program Leadership vs. Program Management Map

FIGURE 9.3
Framework for balancing leadership and management competency.

Scenario 1: Low capability in program leadership and program management. This scenario is not acceptable, as the desired skill set within the program management team is not at the expected level. The people/resource risk is too high for the program to succeed.

Scenario 2: Low capability in program leadership, but high capability in program management. The positive aspect here is that the process, rigor, and discipline in executing and maintaining the transformation program plan through the program management life cycle is well covered. The soft skills to manage the people side and leadership change resiliency are not at the needed level.

Scenario 3: High capability in program leadership and program management. The probability of a business transformation program succeeding is highest in Scenario 3. The program management team is well rounded and possesses the requisite skills to lead the program to success amidst the challenges, issues, change resistance, and risks that will be encountered.

Scenario 4: High capability in program leadership, but low capability in program management. In this scenario, the big picture thinking, change resiliency, and people dimension has a solid coverage in the context of program execution. The hard skills, rigor, and discipline in executing the transformation program plan are not at the expected level.

PROGRAM LEADERSHIP THROUGH OFFICE OF BUSINESS TRANSFORMATION

The leadership needed for complex business transformation programs is optimally provided through centralized centers of excellence, which also facilitate the use of standardized practices and leverage of best practices. In large global enterprises, it is not uncommon to find multiple such centers of excellence, and in smaller enterprises, they tend to be centralized. The multidimensional program management framework presented in Chapter 1 (Figure 1.1) is needed to solve strategic business problems and/or capitalize on business opportunities. The scope, scale, and nature of the problems being tackled by a transformation program dictate the big picture approach needed to solve the problem.

Analysis of the current state and determination of the future state typically has to be done along the three dimensions of people, process, and technology for most business transformation initiatives, as the program manager is focused on improving business results by shifting to a new operating paradigm. In large enterprises, the methods, processes, systems, tools, expertise, and best practices corresponding to each of these three dimensions get institutionalized through "management offices," another term for *centers of excellence*. As enterprises may be running multiple transformation programs concurrently, the expectation is that any individual program will adhere to the management office's guidelines to minimize program risk, avoid reinvention, and execute the program efficiently.

The program management function within an enterprise sets up the program management office (PMO) for housing and deploying program management methods, processes, systems, tools, expertise, and best practices. The domains of work for the PMO (i.e., the things that a typical PMO does) are outlined in detail in PMI's *Pulse of the Profession: PMO*

Frameworks (2013) research output. Though an enterprise level PMO may exist, a PMO dedicated to the transformation program will provide program management leadership and expertise to the team working on the program. The transformation program PMO provides the overall execution leadership through the integration of the three dimensions into the program management practices. Under the leadership of the transformation program management team, the office of business transformation (OBT) is put together under the transformation program PMO to integrate the dimensions of people, process, and technology on the program. OBT comprises the following:

- *Organization change management office*: Capability to provide *people* leadership and houses change management methods, processes, systems, tools, expertise, and best practices. Organization change management was described in Chapter 7. The three phases are:
  - Envision
  - Plan
  - Execute
- *Business process innovation office*: Capability to provide *process* leadership and houses business process redesign methods, processes, systems, tools, expertise, and best practices. The high level phases to get to the innovated business process in the future state are:
  - Current process analysis
  - Future process blueprinting
  - Process gap closure plan
  - Implementation of future process
- *Technology management office*: Capability to provide *technology* leadership and houses technology management methods, processes, systems, tools, expertise, and best practices. On the information systems side, there are many options these days, including off-the-shelf solutions that could be cloud, hosted, in-house, or custom solutions, and the processes to get them deployed vary. The high level phases of the systems development life cycle are:
  - Requirements elicitation
  - Solution options analysis and decision
  - Solution selection
  - System design
  - Development
  - Testing

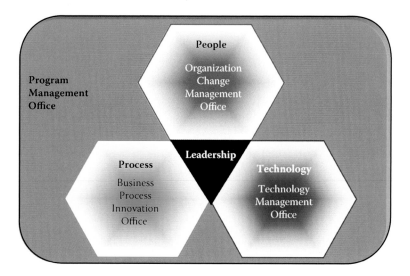

FIGURE 9.4
Office of business transformation.

- Training
- Implementation

Figure 9.4 has a sketch of the office of business transformation.

DRIVERS FOR THE OFFICE OF BUSINESS TRANSFORMATION

There are numerous advantages to setting up such a dedicated office for a large scale multiyear program that is designed to transform the business. Some of the key advantages are listed here:

- Integrated and scalable framework to effectively address the complexity, scope, size, and business impact of the multitude of projects constituting the program
- Increased cross functional and cross project collaboration among the core program team personnel and the opportunity to build trust and strengthen relationships across stakeholders
- Reduced redundant effort by establishing and operating as an integrated team whose charter is to execute an integrated end-to-end transformation plan that supports the shared strategy

- Defined accountability for the timely planning and delivery of the various interdependent work components needed to achieve the program business objectives and business outcomes
- Leveraged learning and roll-out of continuous improvements across projects/components
- Improved operational efficiency in planning and executing the work by capitalizing on standardized approach, repeatable processes, and reusable tools
- Increased organization preparedness for accepting the future state as a result of the structural integration of people, processes, technology, and program management activities

As the transformation program PMO leads the office of business transformation, the PMO's leadership traits play a vital role in helping the organization to implement a transformation program more effectively. In a PMI commissioned research study on strategic PMOs, Forrester Research (2013) found that the PMO leaders recognized that they had to be "part evangelist, part therapist, and part coach" in order to successfully transition the organization to a more disciplined approach to executing a transformation program.

SIGNIFICANCE OF LEADERSHIP IN PROGRAM COMMUNICATIONS

In the discussion on stakeholder expectation management in Chapter 8, the communication life cycle framework was introduced, which is used to drive transformation program messaging throughout the program management life cycle. Transparent, relevant, and aligned messaging by all the leaders on the transformation program positions the program in the right vein in the minds of stakeholders. One of the primary objectives of program level communications is clear, timely, consistent information to the right stakeholders. Another important program communications objective is for the right leader to engage with the right stakeholders at the right time by using the right communication method to build ownership, buy-in, and program acceptance.

Leadership in planning and implementing a robust communications machinery that delivers the appropriate communications objectives will

Communication Planning and Implementation								
Id	Stake-holder/ Audience	Communi-cation Objective	Communi-cation Requirement	Communi-cation Plan	Key Message to Communicate	Communi-cation Method	Communi-cation Owner	Communi-cation Calendar	Expected Outcome
1									
2									
3									
4									
5									
6									

FIGURE 9.5
Leadership in transformation program communications.

go a long way in getting the support of the majority (if not all) of stake-holders on the business transformation charter. The higher the business impacts on stakeholders or the higher the influence of stakeholders on the program, the greater is the need for stakeholder engagement through tai-lored program communications. Communications leadership will mini-mize the speculation around the organization and the individual impacts of the program and avoid derailment of the transformation program's strategic business objectives. Figure 9.5 provides a proven template that can be used to plan and implement program communications and moni-tor the attainment of communications outcomes.

BENEFITS REALIZATION LEADERSHIP

The importance of leadership around the fourteen program management processes was touched upon a bit earlier in this chapter. Business outcome management is one of these fourteen processes that runs end-to-end across the program management life cycle perspective. The business outcome management process will define, plan, execute, monitor, and transition the activities associated with the delivery of program outcomes and real-ization of program benefits. As the ultimate success of a transformation program hinges on the realization and sustainment of the expected ben-efits, leadership of the business outcome management process is critical.

Benefits realization leadership will ensure alignment on expected benefits, provide the needed transparency to stakeholders on benefits-realization status, garner continued support of stakeholders, and keep the core program team motivated. Leadership involvement, oversight, and ongoing support over benefits realization definition, sponsorship, governance, planning, and monitoring are needed. The definition and delivery of objectives, outcomes, and benefits may necessitate negotiating with numerous stakeholders. Experienced leaders possess the soft and hard skills needed to prioritize objectives, outcomes, and benefits. The disciplined management and leadership of the benefits realization related activities by the program management team are a strategic imperative.

Case Study: Business Transformation Initiative on Privacy and Compliance

CONTEXT

A global computer networking company with hardware and software solutions wanted to be the first market mover in embedding privacy and security protection in the solutions it offered to customers.

BUSINESS PROBLEM OR OPPORTUNITY

The well established and publicly traded company with a matured product line had risk exposure from privacy vulnerabilities. The company did not have a dedicated organization that took ownership of privacy matters, including rapid response to security breaches that exposed private information. The primary sources of risk exposure to the company in the current state were the inadequate privacy policies and inconsistent privacy protection practices.

The uncertainty on the definition and evolution of consumer privacy regulations in the United States and how the U.S. regulations would compare to those of other countries posed a challenge in determining which technology standard to support. The ever-changing technology landscape with constant influx of new products that protect privacy as well as create privacy vulnerability added to the threat faced by the computer networking company. The company was keen on exploring whether a good market opportunity existed for it to develop and offer a stand-alone consumer privacy protection product.

SOLUTION

The team contributing to the privacy protection initiative established an end-to-end privacy compliance program. The role of "Chief Privacy

Officer" was defined, and this role was the single point of account-ability for all privacy matters. The program team developed a mul-tilevel leadership model to facilitate planning and implementation of rapid risk reduction projects to counter privacy vulnerabilities. As the privacy compliance program was designed to address privacy from multiple perspectives (strategy, people, process, technology, structure, and measurement), a multidimensional leadership model was implemented.

The lack of a compliance culture at the computer networking com-pany required highly visible leadership for the business to transform to mitigate risk of privacy vulnerability as well as seriously explore if a market opportunity truly existed for the company. The leadership in the planning stages of this business transformation resulted in the development of the company's privacy policy, privacy protection strat-egy, privacy compliance road map, and privacy program communica-tions plan.

BUSINESS OUTCOMES AND BENEFITS

The creation of a dedicated function to take care of privacy matters was a significant business outcome for the company. An intangible benefit was the organization's increased awareness of risk exposure from pri-vacy vulnerabilities. The steps taken by the company to develop and implement a privacy policy and privacy protection practices mitigated the risks. The highest level leadership commitment and leadership at multiple lower levels resulted in the successful launch of the privacy protection initiative at the computer networking company.

SUMMARY

The complexity of business transformation programs requires cross functional, cross dimensional, and cross project leadership to realize the program vision and sustain business outcomes. The significant business change initiated by a transformation program heavily impacts a large number of stakeholders at different levels and affects the organization as a whole. Leadership in analyzing and communicating the impact, rein-forcing commitment to the needed support, and sharing the benefits of transformation program is paramount. Leadership over the development, implementation, and communication of a comprehensive support strategy to counter the change impact is essential to increase the acceptance rate of the program among the impacted parties.

Leadership behavior has to permeate through multiple levels of the transformation program organization and not just at the level of the executive sponsor or the steering committee. Leadership on the technical and human side is essential for the entire duration of the program management life cycle. The office of business transformation provides an integrated leadership model that facilitates getting the right quantum of leadership to all of the dimensions of program management. Program leadership—or a lack of it—can make or break a program devised to transform a business.

REFERENCES

PMI. 2013. *Pulse of the profession in-depth report: PMO frameworks.* Newtown Square, PA: Project Management Institute.

Forrester Research. 2013. *Strategic PMOs play a vital role in driving business outcomes.* Newtown Square, PA: Project Management Institute. http://www.pmi.org/~/media/PDF/Publications/Forrester-PMOs-Play-Vital-Role-TLP-PMI-Final.ashx

10

Monitor Aggressively and Have Contingencies

The program management team must develop and implement rigorous processes and tools to closely track progress of the business transformation program. Similar to an automobile that fails if routine oil changes are ignored, ongoing and timely corrective action is critical to avoiding major setbacks to the mission, vision, and objectives of the transformation program. Throughout the business transformation journey, checks and balances are necessary to not only ensure that the program is advancing as envisioned, but also to correct course on a timely basis. The "monitor program delivery" process of our eight step program management cycle is designed to keep a close view on the program. The "monitor program delivery" process and the "execute program" process are happening concurrently.

The outputs generated by the program need to be assessed on an ongoing basis to validate that they are in line with the specifications and stakeholder expectations, and those that are not will need to be rectified. Even the most well planned and rightly resourced programs need to have a contingency plan in place that can be used to handle anything unexpected that could affect the program during any stage of the program management life cycle. The contingency plan has to be tested, and the program manager must be continually thinking ahead in terms of a "What is our plan B?" mindset. This chapter presents best practice templates as well as techniques that can assist with program monitoring. The use of a program governance model and the implementation of governance processes also play key roles in controlling the program.

The following topics are described in this chapter with the help of supportive illustrations. including a real world case study:

- Monitoring status of transformation program
- Program monitoring: key criteria
- Transformation program status dashboard
- Tracking and reporting project performance
- Monitoring of transformation program: milestones, financials, issues and risks, and change requests
- Enabling continuous improvement
- Contingency strategy and contingency planning
- Monitoring benefits realization
- Case study: program to manage transformation of business processes

MONITORING STATUS OF TRANSFORMATION PROGRAM

All program stakeholders need to be kept apprised of how the program is coming along in delivering against the transformation program plan. On the tactical side, the program manager sets up and implements the regular ongoing meeting forums to share program status. On the strategic side, the governance forums are used to review and monitor program health. A sample of governance bodies and governance rhythm from a transformation program is shown in Figure 5.3 in Chapter 5. The program management team devises and implements program status development, dissemination, and archival processes early on in the program and communicates those to the stakeholders. Program status reports take the form of high level program health dashboards and granular program status reports. The content within these status reports and their structure is tailored to meet a business transformation program's stakeholder needs as well as the stage of the transformation program. Based on the objectives, attendees, timing, and frequency of the governance forums, the relevant types of program status reporting materials are developed and shared prior to the meeting and then reviewed during the meeting. Since a large program is highly likely to have many stakeholders, some of them may not be attending any of the ongoing status meeting forums, and any specific program health related inquiry or concern they may have will get addressed though their respective points of contact per the program's organizational structure and the program's governance model.

A business transformation program typically comprises multiple projects spanning multiple dimensions (i.e., people, process, technology, etc.) and impacted functional areas, and these interdependent projects are integrated and managed together with the help of the program plan to achieve efficiencies. Based on the nature of the program and what it is designed to accomplish, some projects will naturally come to an end even before the program is formally complete, and other projects will get launched midstream of the program management life cycle. The overall status of the transformation program is influenced by the status of the various individual projects.

The top half of Figure 10.1 is constructed to determine the status of each project, which can be done with the help of the seven criteria in the first column: timeline, funding, resources, scope, quality, risk, and value. The next three columns specify the basis to determine the green or yellow or red status for each of the seven criteria. For example, as the project is motoring ahead, if the funding is in line with the budget, then the "financial" criterion for the project is green. If the project may need extra funding, the project is yellow on the "financial" criterion, and if the burn rate has exceeded the budget or will exceed for certain, the project is red on the "financial" criterion. The last row of the top table provides the aggregate

Project status criteria and definitions			
Project status criteria	**Green**	**Yellow**	**Red**
Timeline	On track	Milestone may be missed if corrective action is not taken	At least one milestone has been missed
Funding	Within budget	Additional funds may be required to meet project success criteria	More funds needed to meet project success criteria
Resources	Fully staffed	Resource capacity constrained	Additional resources need to be procured
Scope	Aligned to program charter, project boundaries and plan	Scope may get out of alignment due to changes	Not aligned as needs have shifted
Quality	In line with specifications	Some specifications are likely to be missed	Some specifications have been missed
Risk	Mitigation plan is aligned to risk level and trigger probability	Mitigation plan may not be aligned to risk level and trigger probability	Mitigation plan is not aligned to risk level and trigger probability
Value	On track to deliver the business benefits based on KVIs	Delivery of business benefits may be missed based on KVIs	At least one key expected business benefit will not be delivered
Overall	**All vitals are GREEN**	**None of the vitals are RED and at least one is YELLOW**	**At least one vital is RED**

Program status criteria			
Program status criteria	**Green**	**Yellow**	**Red**
Overall	**All projects are GREEN**	**None of the projects are RED and at least one is YELLOW**	**At least one project is RED**

FIGURE 10.1
Transformation program monitoring: key criteria.

basis to determine whether the project is in green or yellow or red status. If the project has green status for all of the seven criteria, the overall health of the project is green. If one or more of the seven criteria for the project is in yellow status, the overall project health is yellow. If one or more of the seven criteria for the project is in red status, the overall project health is deemed red.

The bottom half of Figure 10.1 is constructed to determine the status of the transformation program. The overall program status or program health is driven by the status or health of the individual projects. Hence, each project needs to be well managed and controlled. To monitor program status rigorously and holistically, the status of each project is analyzed, tracked, and reported against these seven criteria. The last row of the second table in Figure 10.1 provides the basis to determine whether the status of the program health is green or yellow or red.

For example, if all of the projects constituting the program have a green status, the program status is green. If one or more of these projects is in yellow status, the program health is yellow. If one or more of these projects is in red status, the program is in red status. The progress and status of a transformation program should be monitored aggressively by capturing, analyzing, and reporting status on a weekly basis and taking the necessary actions at the program and project levels to constantly maintain good program health.

PROGRAM MONITORING: KEY CRITERIA

Figure 10.2 highlights the application of the project status criteria and program status criteria described in the prior section with the help of an example. In the example, the program subsumes four projects, with the health of each project captured in the "overall" column. The first project has green status for the six criteria of timeline, funding, resources, scope, quality, and risk and yellow status for the seventh criterion of value, which leads to an overall yellow status for the project. The second project in the program has a yellow status for the resources criterion and a green status for the other six criteria, which also results in a yellow overall status for that project. The third project is marching along well with an overall green status, as all of the seven status criteria are green for the project. The fourth project within the program is also firing on all cylinders, with an

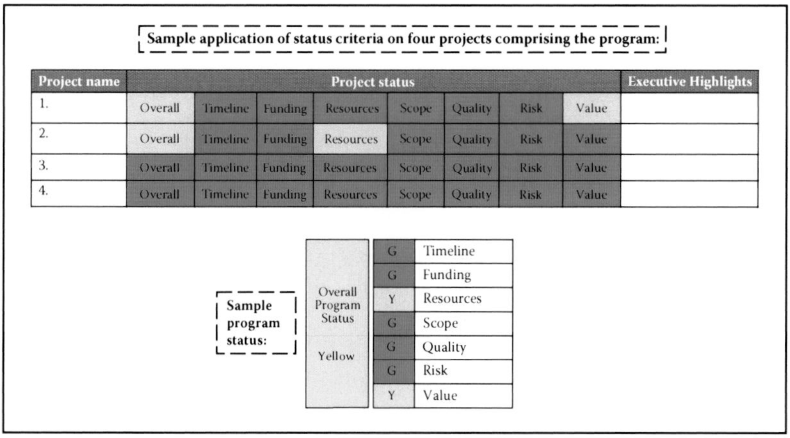

Project name	Project status								Executive Highlights
1.	Overall	Timeline	Funding	Resources	Scope	Quality	Risk	Value	
2.	Overall	Timeline	Funding	Resources	Scope	Quality	Risk	Value	
3.	Overall	Timeline	Funding	Resources	Scope	Quality	Risk	Value	
4.	Overall	Timeline	Funding	Resources	Scope	Quality	Risk	Value	

Sample application of status criteria on four projects comprising the program:

	G	Timeline
	G	Funding
Overall Program Status	Y	Resources
	G	Scope
	G	Quality
Yellow	G	Risk
	Y	Value

Sample program status:

FIGURE 10.2
Transformation program monitoring: application of key criteria.

overall green status. With the first two projects in yellow status and the next two in green status, the overall status of the program is yellow.

The status reporting tool in Figure 10.2 is effective in capturing and reporting status to stakeholders and in managing stakeholder expectations along the way. The tool can be enhanced and tailored to address the differing needs of different stakeholder types. However, the program management team needs to ensure that the status report tool doesn't get too complicated, as that would reduce the user friendliness and increase the effort level needed to operationalize the tool. The recommended monitoring frequency is weekly, and periodic analysis of such weekly program status reports for a period of time will provide the program management team with invaluable insight about the program trends. The trending information provides lagging and leading indicators to the program manager in terms of what remediation actions are working and what new actions need to be taken to bring the program health back to green status and reposition it to realize the desired business outcomes.

TRANSFORMATION PROGRAM STATUS DASHBOARD

The execution and monitoring of the business transformation program go hand in hand, and both are part of the "make it happen" stage of the program management life cycle. The program management team devises and implements various processes and sub-processes to constantly monitor the execution status and outputs coming out of the transformation program and initiate timely corrective actions. In Chapter 7, fourteen program management processes were touched upon, with some of these process examples being stakeholder management, governance management, integration management, scope management, time management, cost management, and quality management.

At a high level, the business transformation program's monitoring function covers all bases by having monitoring activities pertaining to each of these fourteen program management processes. The governance and change control processes fall under the monitoring function, and these ensure that any modifications to the baselined program deliverables (e.g., program road map, program charter, program plan) and tangible outputs are allowed only after those have gone through the established change control protocol. The controls that are institutionalized on the program by the monitoring activities enable efficient, optimal execution of the program and prevent rework, which consumes additional unplanned resources. A tight, close, and constant monitoring of the program is mandatory to manage the complexity and strategic risks inherent in a program designed to transform the entire business or significant chunks of the business.

The communications management processes are responsible not only for capturing and distributing information pertaining to the outputs and outcomes coming out of the business transformation program to the stakeholders, but also for sharing how the program is performing in relation to the integrated transformation program plan. The framework to assess how the program is performing and for reporting program performance in the form of status reports was covered in the preceding section. The ownership of reporting the program's status falls under the program communications team, and program status could be reflected in different types of artifacts, with the most common being weekly program status reports and program dashboards. Typically, program status reports are comprehensive, as they capture granular information, and program dashboards are higher level roll-ups. Both of these program performance reports are needed, as they cater to different audiences.

The chart in Figure 10.3 accentuates a sample program status dashboard, which is essentially a composition of the following building blocks:

Overall Program Status			Project name	Project status							
	G	Timeline									
	G	Funding	1.	Overall	Timeline	Funding	Resources	Scope	Quality	Risk	Value
	Y	Resources	2.	Overall	Timeline	Funding	Resources	Scope	Quality	Risk	Value
	G	Scope	3.	Overall	Timeline	Funding	Resources	Scope	Quality	Risk	Value
Yellow	G	Quality	4.	Overall	Timeline	Funding	Resources	Scope	Quality	Risk	Value
	G	Risk	5.	Overall	Timeline	Funding	Resources	Scope	Quality	Risk	Value
	Y	Value									

Program executive summary	Key program deliverbles/milestones	Due date	Trend	Owner

Key program accomplishments for last week	Key program activities for this week	Key program decisions

Major program issues and risks	Major program change requests	Major program dependencies

FIGURE 10.3

Transformation program status dashboard.

- Overall program status
- Overall summary status of each project comprising the program
- Program executive summary
- Key program deliverables and milestones
- Key program accomplishments for the last week
- Key program activities for this week
- Key program decisions
- Major program issues and risks
- Major program change requests
- Major program dependencies

These building blocks can be tailored based on the needs and preferences of the program sponsors, steering committee, executives, governance body representatives, and other stakeholders. Program management determines the composition of the dashboard based on stakeholder needs, maps out the dashboard development steps, and then implements them. The end goal is to keep the program status dashboard—and the process to develop it—as simple as possible.

TRACKING AND REPORTING PROJECT PERFORMANCE

A business transformation program drives transformational change by revamping multiple segments of the current state of the business to the envisioned future state. To bring such a renewal to fruition by reaching the future state—and with stakeholders embracing the future state—the transformation program has to design an integrated plan that cohesively brings together the different segments. The entire body of work to be accomplished by the program is modularized within the integrated program plan by grouping logical units of work together. These modules can be planned and executed on a stand alone basis, and their interdependencies are factored into the integrated plan. In program management terminology, these modules are referred to as *projects*. Program monitoring entails monitoring these modules, as program performance is influenced and dictated by the performance of the modules/projects that make the program.

The project management structure within the program organization model is accountable for the planning, execution, and delivery of project level outcomes. Project status can be captured and communicated with the help of a project status dashboard, which is spotlighted in Figure 10.4. It typically includes the following:

Project Manager	Project Name		Project Status							
		Overall	Timeline	Funding	Resources	Scope	Quality	Risk	Value

Project executive summary	Key project deliverables/milestones	Due date	Trend	Owner

Key project accomplishments for last week	Key project activities for this week	Key project decisions

Major project issues and risks	Major project change requests	Major project dependencies

FIGURE 10.4
Diligent tracking of projects comprising transformation program.

- Overall project status
- Project executive summary
- Key project deliverables and milestones
- Key project accomplishments for last week
- Key project activities for this week
- Key project decisions
- Major project issues and risks
- Major project change requests
- Major project dependencies

MONITORING TRANSFORMATION PROGRAM

Milestones

Milestones are significant events on the business transformation journey, and these are mapped out as part of an integrated program plan development process. Examples of milestones include completion of the program road map, release of funding, completion of program planning phase, and completion of systems design. As the program plan is a roll-up of multiple project plans, the key project level milestones from the project plan are reflected in the program plan. Since a typical business transformation program plan can run into multiple pages, it's a good practice to abstract all the milestones in a separate simple output (see Figure 10.5) and track their status in that output. Diligence in tracking milestones and sharing their status is critical, as push-outs in the date of completion can have a negative impact on the critical path of the transformation program, which can delay the overall timeline.

Maintenance of the program plan by the program management team can result in new work getting added, planned work getting removed, date shifts occurring, etc., and such changes can influence program milestones. The following elements have to be analyzed and reported regularly to program stakeholders as part of the milestone monitoring process:

- Milestone (or event) description
- Milestone status (green, yellow, or red) based on work progression
- Impact to program from yellow and red milestones
- Actions and action owners for yellow and red milestone
- Planned completion date for milestone

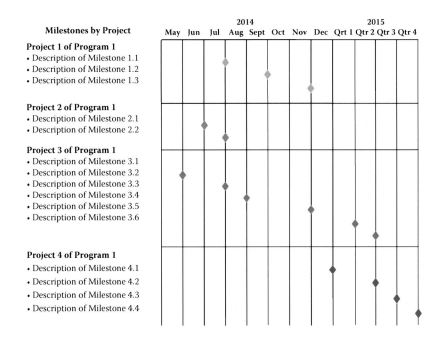

FIGURE 10.5
Monitoring transformation program milestones.

- Actual completion date for milestone
- Explanation of variation

Financials

A program designed to transform a business has to attain that end state within the financial parameters. The program management team has to manage and watch the actual costs of the program, and the costs can fall under different categories. At a high level, there are business and technology cost categories, and the technology costs can be classified as operational or capital expenses. Operational expenses include the ongoing costs for running the business and the supporting technology. Capital expenses include fixed asset purchases (e.g., computer hardware, software) and the labor expenses to install them, and these create future benefits.

Figure 10.6 displays a sample deliverable from the ongoing program financial monitoring process. Actual program expenses are tracked against the budget over different periods of time, and the reasons for variances are noted. Given the typical multiyear duration of a business

Business Costs ($ in millions)	Current Year Estimate:				Overall Program:			
	2014 Budget To-Date	2014 Spend To-Date	Variance	Variance Explanations	Planned Budget To-Date	Actual Spend To-Date	Variance	Variance Explanations
Total Expense								
	Budget	Actual						
Last Month								
Last Quarter								
Year To Date								

Information Technology Costs ($ in millions)								
	Current Year Estimate:				Overall Program:			
	2014 Budget To-Date	2014 Spend To-Date	Variance	Variance Explanations	Planned Budget To-Date	Actual Spend To-Date	Variance	Variance Explanations
Total Expense								
1. Operating Expense				Operating Expense				
2. Capital Expense				Capital Expense				
	Budget	Actual						
Last month								
Last Quarter								
Year To Date								
Total Costs								

FIGURE 10.6
Monitoring program financials.

transformation program, financial information for each year and for the program to date is maintained and tracked through financial monitoring. The transformation program sponsorship team needs details of program performance from a financial perspective, and this is usually shared by the program manager via the governance forums devised for the program.

Issues and Risks

Due to the complexity and significant change a business transformation is driving, there are significant barriers to the business transformation program attaining the future state blueprint and sustaining the future state. The program management team should have a systematic approach to handling program issues (active problem) and program risks (potential problem). The top half of Figure 10.7 exhibits a sample program issues management tool, and the bottom half exhibits a sample program risk

| | | | | | | Program Issues Management Tool | | | | |

Id	Issue Description	Impact on Timeline, Funding, Resources, Scope, Quality, Risk, and Value	Issue Criticality (high, medium, low)	Resolution Plan	Due Date	Owner	Escalate /Inform	Status (open, closed)
1								
2								

Id	Risk Description	Potential Impact on Timeline, Funding, Resources, Scope, Quality, Risk, and Value	Probability	Impact	Risk Trigger (conditions to monitor)	Risk Severity (high, medium, low)	Risk Mitigation Plan	Due Date	Owner	Escalate /Inform	Status (open, closed, watch list)
1											
2											

FIGURE 10.7

Disciplined program monitoring results in timely action.

management tool. It's a common industry practice to refer to these tools as *program issues logs* and *program risks logs*, which are outputs of the program issues and risk management processes. Project managers own the project-level issues and risks, which impact only the project and not the entire program, and these are governed by the project level management processes for issues and risks. The benefits of having well-oiled management processes for issues and risks at the program level as well as the project level cannot be overestimated.

Program managers own the program level issues and risk management cycles from beginning to end. The issue management cycle includes issue identification, issue analysis, issue resolution, issue escalation, issue tracking, and issue reporting. The criticality of an issue is based on the impact it is having on the program timeline, funding, resources, scope, and quality. If resolution to any issue necessitates changes to what has been planned and baselined, the changes have to go through the program governance process of program change control. The end-to-end risk management cycle encompasses risk identification, occurrence anticipation, impact analysis, risk mitigation, risk escalation, risk tracking, and risk reporting. The program management team defines and communicates the escalation criteria and escalation steps. A lack of tenacity in managing program level issues and risks has severe consequences on the business transformation program.

Change Requests

The forward progression and health of the business transformation program is tracked and reported to key stakeholders on a weekly basis. The tools of program status dashboard (Figure 10.3) and project level tracking (Figure 10.4) are utilized for the same. Significant program level or project level changes to the baseline that has been signed off by all stakeholders can negatively impact one or more of the seven criteria (timeline, funding, resources, scope, quality, risk, and value) used to assess the overall program health and the heath of the projects constituting the program. As a result, the core program team has to maintain a constant vigil on the program level and project level change requests that get raised. A change request that potentially jeopardizes the delivery of the business outcome and benefits expected by the stakeholders is deemed as a critical or major change request. The weekly program level and project level dashboards specify the major change requests and the actions being taken on them.

The program manager has to ensure all the change requests are being scrutinized through the defined change control process. In expounding the program governance model in Chapter 5, the importance of the change control process was highlighted. Figure 5.5 presented the framework to capture, analyze, and review the critical change requests. The change control process will have escalation paths and decision making guidelines on the approval or denial of change requests. The program management team has to ensure the program governance on change control management is adhered to and disallow any exceptions to the signed-off change control process. The monitoring of program level and project level change requests by the program management team is crucial.

ENABLING CONTINUOUS IMPROVEMENT

Monitoring the processes of a business transformation program provides a platform for creation of a continuous improvement environment. Monitoring activity happening during the "make it happen" phase of the program management life cycle detects active problems and warning signs of potential problems. The program management team devises and implements corrective action plans to address the root causes of issues unearthed through continuous monitoring. Monitoring and controlling the program generates a feedback loop that can be harnessed beyond the immediate corrective action. The facts and insights collected from the feedback loop can be used to drive step change improvements within the prior program management life cycle phases of "set the stage" and "decide what to do."

Application of the tool portrayed in Figure 10.8 is one of the ways to get additional strategic benefits from monitoring of a transformation program. Under the leadership of the core program management team, the voices of the stakeholders and subject matter experts can be gathered and documented along the themes of "What's working?" "What's not working?" and "What's the recommendation?" The performance of such an exercise on a quarterly frequency across the different segments (or projects) of the business transformation program provides invaluable nuggets of knowledge. An analysis of the captured collective wisdom from the various stakeholders and business areas engenders continuous improvement ideas and recommendations that could be deployed on the transformation program in progress as well as on future transformation programs after socialization and approval. Such deployments instill a continuous improvement mindset within the program team, create a sense of ownership, and pave the path for organizational acceptance of a program.

Program Name:		Program Sponsor:	
Project/Track Name:		Project/Track Manager:	
Period Covered:		Program Manager:	

Id	What's working? (Strengths)	What's not working? (Challenges)	What's your recommendation? (Improvements)
1			
2			
3			

FIGURE 10.8
Creating a continuous improvement program culture.

CONTINGENCY STRATEGY
AND CONTINGENCY PLANNING

One of the sources of complexity in a transformation program is the broader and deeper impact to the organization due to the numerous interdependencies. As discussed previously, program monitoring assists with keeping a close eye on the progress of executing the transformation program. Monitoring involves assessing program performance at any point in time by comparing the actual against the plan from the perspective of a timeline, funding, resource allocation, scope, quality, risks and value. Program sunk costs (incurred cost) and program opportunity costs (immediate cost of doing some other program) for large scale programs driving significant business change at an enterprise or business unit level is high.

Mitigation of risks is an important critical success factor for a business transformation program, as they typically consume a large quantum of organizational resources for an extended period of time. Mitigating high and medium severity risks is imperative in the case of a business transformation program. Per the PMI's *Pulse of the Profession In-Depth Report: Organizational Agility* study (2012) that was conducted among 1,239 practitioners, organizations have increased their success rates on initiatives or programs by implementing risk management best practices that included formal contingency planning. The risks encountered by a transformation program can be mitigated by having a well thought out contingency strategy and contingency plan in place, which will drive the right set of actions at the right juncture.

Contingency strategy and contingency planning are the front end components of the end-to-end risk management cycle. Contingency strategy needs to factor the complexity by having a fallback plan, which would kick in if the pretested and working contingency plan is unable to achieve all of the plan objectives for any reason. The contingency strategy for the transformation program also has to be designed to address any new risks that might be introduced from triggering of the contingency plan and fallback plan. It is vital to test the contingency plan in advance, as this will ensure that the contingency plan will achieve the intended objectives when the identified risk is triggered. The program management team should have buy-in of the executive sponsor and steering committee for investing in contingency strategy and planning. Even if the needed investment for contingency planning is high, it is important to recognize that the tangible and intangible return on that investment is also high.

MONITORING BENEFITS REALIZATION

The program monitoring function involves maintaining a constant watch as the transformation program plan is executed by the program team. The benefits realization plan is a component of the transformation program plan. The objective of benefit realization monitoring is to ensure that the delivery of desired outcomes and the realization of expected benefits will be in line with the benefits realization plan. The monitoring of benefits realization identifies roadblocks, issues, and risks that specifically impact the delivery of business outcomes and realization of benefits. In addition, benefits realization monitoring devises and implements action plans to overcome the identified roadblocks and issues.

The execution of the following benefits realization monitoring activities plays a significant role in the transformation program attaining the desired future state, accomplishing the strategic objectives, and sustaining the realized benefits in line with the plan:

- Assessing work progress against the benefits realization plan
- Taking timely corrective actions if work is not progressing per the benefits realization plan
- Reporting the status of execution of the benefits realization plan to keep stakeholders informed
- Raising change requests to get approval on deviations from expected benefits
- Communicating deviations in benefits to stakeholders to reset expectations
- Resolving issues impacting the realization and sustainment of benefits
- Identifying and mitigating risks pertaining to realization of benefits
- Devising contingency plans to deliver outcomes and benefits

---

Case Study: Program to Manage
Transformation of Business Processes

CONTEXT

A highly successful health care organization had been developing and launching new health plan products to cater to different market segments in the United States. After substantial market research, the organization launched a new health plan product offering with a much higher deductible but a much lower insurance premium in relation to its existing health plan products. The product was a great success and started adding new members at a very fast clip. Unfortunately, the business operating platform and business readiness to support the administration of this product could not scale, and this impacted the satisfaction of external stakeholders, i.e., members, brokers, and customers.

BUSINESS PROBLEM OR OPPORTUNITY

Given the external and internal environments, the health care organization had to expeditiously overcome the challenges. They ran the risk of losing existing members, experiencing slower growth in the product, and suffering revenue leakage due to improper administration. There was no organized and coordinated effort to manage and monitor the postproduct launch activities. The business operations platform supporting the administration of the new product had capability gaps. The gaps in administration processes, customer support processes, and policies needed to be addressed effectively. The lack of contingency plans and workarounds resulted in poor response time to the members with a resolution to their issues.

SOLUTION

The charter of the business transformation program in place to launch the new health plan product was expanded to handle post product launch support. The program management team worked with stakeholders to build a program plan to address the identified challenges and manage the postlaunch phase. The business process improvement and policy enhancement work commenced. The transformation program team implemented standardized program management processes to track status, issues, risks, changes, and financials. In addition to the detailed weekly status reports, a high level biweekly program status dashboard was put in place to manage the expectations of stakeholders at different levels. In monitoring the execution of all the work captured in the program plan, the program manager ensured enhancement of the existing training, support, and contingency planning processes. The program manager solicited inputs of external and internal stakeholders on continuous improvement ideas to the processes and policies, which would enable the health care organization to deliver the highest level satisfaction.

BUSINESS OUTCOMES AND BENEFITS

The redesigned business processes to administer the new product with the help of trained, experienced personnel led to better member and customer support through faster response times. The customer experience in dealing with the product administration function got better, and the warning signs of potential customer attrition went away. The risk exposure to the health care organization was minimized with the policy enhancements, renewed processes enforcing compliance to policies, and a contingency plan. The roll-out of the continuous improvement plan further improved the readiness. The aggressive monitoring of the program execution, including the placement of a contingency plan, enabled the health care organization to protect its market reputation.

SUMMARY

Institutionalization of program controls is essential for efficient operational execution of the business transformation program and to position the program for success. The program management life cycle process of "monitor program delivery" interjects these controls and assists with effective management of the strategic and tactical risks to the business transformation program. The timely and accurate capture, analysis, and communication of the program status to the program stakeholders aids in timely interventions for the program to stay aligned to the plan. Actions needed to correct the course are determined by the monitoring processes, and implementation of the approved actions happens through the program execution processes.

A transformation program status dashboard and project level dashboards are used to keep stakeholders informed of the program's status and meet stakeholder expectations. Monitoring of program milestones validates that the program timeline is in line with the integrated program plan. Program scope validation confirms that the produced outputs are in line with the charter, and program financial tracking ensures that budget parameters are met. The issues and risk management tools support the implementation of the critical issues and risk management processes. The program monitoring process drives additional value into the business transformation program by creating opportunities for continuous improvements. Finally, tested contingency plans mitigate the significant risks encountered by the transformation program.

REFERENCE

PMI. 2012. *Pulse of the profession in-depth report: Organizational agility.* Newtown Square, PA: Project Management Institute. http://www.pmi.org/~/media/PDF/Research/Organizational-Agility-In-Depth-Report.ashx

11

Create and Implement an Operations Transition Plan

The significant business changes being driven by a complex, cross functional program (e.g., to plan, design, and implement a new product, service, or system) need to be ultimately absorbed by the pertinent operational functions within the enterprise. A transformation program transition model is codeveloped by working closely with the program stakeholders in the operational groups. An end-to-end program transition plan constitutes the operations plan, training plan, and knowledge transfer plan. Through a well designed and tested program transition plan, the program management team prepares the operations function to handle the future state delivered by the business transformation program.

The program transition plan supporting the transition model has to be built in advance to allow enough lead time for the operations function to be prepared to accept this change. As the transformation program begins to approach completion, a smooth transition from the core program team to the operations team happens. The "transition to operations and close program" process of the program management life cycle is designed to cover all the bases to ensure a smooth transition.

The following topics are described in this chapter with the help of supportive illustrations, including a real world case study:

- Operations transition framework
- Transition model development
- Training the operations team
- Transition model execution
- Value enhancement analysis technique: Overview, objective, and approach
- Transition model operationalization
- Walk through technique: Overview, objective, and approach

- Leading lessons learned from walk throughs
- Business outcome delivery and sustainment
- Case study: Strategic initiative to get and stay compliant with government regulations

Once the program officially closes, the operations group owns the realization of the business objectives and desired business outcomes. Operations readiness entails not only the capability to deliver the outcomes in the near term upon program completion, but to also sustain the delivery of outcomes over time. To do so, the business change has to be accepted and embraced by the operations organization once the transformation program ends. Hence, planning and executing on the "people" dimension with the application of the organization change management (OCM) techniques is a critical success factor. The transition management strategy and transition management plan covered under the OCM framework (Chapter 7) is a subset of the overall body of work needed for a successful, seamless transition to operations.

Ideally, testing of the program transition plan through a well chosen pilot endeavor should be done prior to the complete switchover from core program team to operations team. The planning and implementation of the program transition plan improves the comfort level, enables organizational buy-in, and empowers the operations team. Finally, the planning and roll-out of a formal knowledge transfer process facilitates a smooth transition. A well orchestrated transition of the program ensures that the organization's performance is at the desired level, with the external and internal customers of the program getting top tier support from the operational units after the business transformation program team has been formally dissolved.

OPERATIONS TRANSITION FRAMEWORK

Answers to the following questions assist in formulating, implementing, and sustaining an end-to-end framework that facilitates an effective, efficient, and smooth transition of the program:

- What's changing in moving from the current state to the future state?
- Who is getting impacted by the change, or who can impact the desired business outcome?

- How will the impacted business functions and stakeholders be prepared?
- What is required to sustain the future state and realize the targeted business outcomes?
- How will business performance levels in the future transitioned state be monitored?

The delivery of the desired transformed future state by the program is just winning half the battle. The other half of the battle is won by sustaining the future state, which ultimately results in the ongoing realization of the defined business outcomes and benefits.

The PMI's *Pulse of the Profession In-Depth Report: Enabling Organizational Change through Strategic Initiatives* (2014) highlights that programs (or strategic initiatives) have to manage people through the change of current state to future state in order to create sustainable change. The program management team can utilize the operations transition framework to effectively manage stakeholders through the change and create a sustainable future state. Figure 11.1 depicts a three-phase approach to transition to the operations group. The phases are as follows:

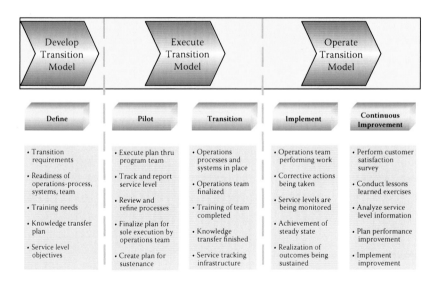

FIGURE 11.1
Operations transition framework.

- **Transition model development:** The transition model has to be devised from a strategy, people, process, technology structure, and measurement perspective. The transition strategy will drive the transition organization, processes, and supporting systems. The transition requirements are defined, documented, vetted, and agreed upon, and the transition model to meet these requirements is constructed. The key elements of the transition model are the operations plan, training plan, and knowledge transfer plan, and these will enable transition readiness of business and technology.

- **Transition model execution:** In the "transition model execution" phase, the actual transition from the program to ongoing operations happens. The prior phase ensured that the operational units are ready. In this phase, the implementation of the transition plan (i.e., operations plan, training plan, and knowledge transfer plan) occurs. A business performance or service level monitoring infrastructure is set up. To mitigate any major risks originating from the transition to operations, the transition plan is tested by the program management team with the help of a pilot program. The plan, process, and systems are adjusted as needed based on the findings and insight from the pilot program.

- **Transition model operationalization:** The "transition model operationalization" phase is an ongoing one, as it pertains to the business delivering the goods and services to customers. The future state attained upon completion of the business transformation program is being executed by the operations team. The business objectives are getting accomplished, and business outcomes are being delivered. Proactive and reactive corrective action plans are being implemented. To constantly raise the bar and sustain the business results delivery, a continuous improvement culture and mindset is created. Results of administered customer satisfaction surveys and ideas from formal lessons learned exercises assist with the creation of an improvement plan. Implementation of improvements sustains the expected business performance.

TRANSITION MODEL DEVELOPMENT

One of the key elements of the transition model is the operational plan, which drives the readiness of the operational process, systems, and team. The change readiness model, readiness strategy, and readiness plan highlighted under the OCM framework (Chapter 7) address the overall organizational readiness from a people or team perspective. Planning and execution of the change readiness activities for operations facilitate acceptance by operations units. The readiness of operations units is one of the critical success factors for the business transformation program to realize and sustain the defined business outcomes. In light of the complexity and disruptive change driven by the typical complex transformation program, understanding of the aggregate impact on the operations and transition requirements is critical.

Figure 11.2 illustrates the proven template that can be put to use to plan and implement the steps needed for the operations functions to be fully ready by the time the program officially closes. The change readiness objective and granular change readiness requirements are elicited for the operations audience. The operations plan (or operations change readiness plan) to fulfill the transition objective and meet transition requirements is then put together. The focus of the operations change readiness plan is on the operations functions only, and there will be other change readiness plans for other impacted organizational groups. All the readiness plans are aligned and integrated into the integrated program plan for the business transformation endeavor. In working with the program sponsorship and leadership teams, the program management team lines up the ownership for readiness of operations functions and communicates that to the stakeholders. The key operations change readiness events can be captured in a calendar view to facilitate organizational awareness. From a due diligence perspective, an institutionalized process for tracking the realization of expected change readiness outcomes of the operations readiness

Change Readiness Planning and Implementation							
Id	Stake-holder/ Audience	Change Readiness Objective	Change Readiness Requirement	Change Readiness Plan	Change Readiness Owner	Change Readiness Calendar	Expected Outcome
1							
2							
3							
4							
5							
6							

FIGURE 11.2
Transition planning and execution: program change readiness.

activities should be implemented. The operations change readiness plan needs to promptly address identified variances.

TRAINING THE OPERATIONS TEAM

Programs designed to transform a business drive significant changes to the current practices followed by the operational functions. For the operations organization to accommodate and adjust to the changes, training is essential. In Chapter 8, training was elaborated as a good vehicle to engage and manage the expectations of stakeholders. The impacted operational groups are one set of key stakeholders. The training life cycle of training strategy through training implementation was introduced in that chapter. Training has to be tailored by the impacted stakeholder group based on the change impact of the program to that group and the role of that group. For the operations audience, both business and technical training may be needed, and separate customized training plans can be crafted for these two sets of audience.

Figure 11.3 demonstrates the tool that can be used to plan and implement training. After identification of the operational audience, the higher-level training objective and the supporting detailed training requirements to meet the objective are determined. Then a training plan that is aligned to the overall training strategy is constructed. The plan has to be comprehensive to map out all the training topics and the method (or channel) to be used to deliver the training. Examples of training methods would be classroom instructor led, virtual instructor led, web based, self training videos, etc.

Training Planning and Implementation										
Id	Stakeholder/ Audience	Training Objective	Training Requirement	Training Plan	Key Training Topics	Training Method	Training Content Owner	Training Delivery Owner	Training Calendar	Expected Outcome
1										
2										
3										
4										
5										
6										

FIGURE 11.3
Transition planning and execution: program training.

Typically, the training materials development team owns content creation, and the training delivery team delivers the training. The training delivery team provides change inputs to the training content development team based on the feedback provided by the trained audience via the post-training evaluation process. A training calendar, a subset of the training plan, spells out the timeline for materials development as well as delivery. A process that validates that the expected outcome of the operations training is being realized has to be institutionalized. Such a validation will ensure that the readiness of operations is at the expected level, which will enable a smooth transition from the core program team to the operations team.

TRANSITION MODEL EXECUTION

The completion of execution of the transition model positions the operations function to start delivering the objectives and outcomes of the transformation program. Prior to official program closeout, the milestone of "transition model execution completed" has to be realized. From a risk management perspective, the execution of the transition model finalized in the earlier phase is completed by a two-stage iterative process. In the "transition" stage, the defined operational processes are made ready to go, and the operations team is finalized and put in place. All the training and coaching activities are completed. The transfer of knowledge from the core transformation program team to the operations team is completed. In the pilot stage, the operations team executes the program with the as needed support of the program team. The knowledge and inputs gathered from the pilot execution is used to refine process, organization, and system. Also, any gaps identified during the execution of the pilot stage of the transition model should be bridged.

VALUE ENHANCEMENT ANALYSIS TECHNIQUE

Overview

Value enhancement analysis is a technique for decomposing operational processes into activities and analyzing them. Though this technique is covered in this chapter, please note that it could be applied in executing some of the other program management life cycle processes too. Value enhancement analysis is also leveraged in the business process improvement framework to assess and enhance the "process" thread of the business transformation program. All the operational (business and technology oriented) processes need to be periodically analyzed as full blown implementations of the transformed state occur through operational units. There are two kinds of value derived from the activities constituting the operational processes:

- *Real value activity*: This activity directly contributes to meeting the end customer's expectations.
- *Business value activity*: This activity adds no value from the end customer's perspective.

Objective

The purpose of value enhancement analysis is to maximize the value to the end customer of the ultimate output (product, service, process, system, etc.) delivered by the business transformation program. Upon formal closure of the program, the day-to-day operational processes will continue to create, maintain, and enhance the output. Ultimately, "value" must be measured from the perspective of the end customer or consumer. Value is anything the consumer is willing to pay for, whether it is a product feature, the cachet associated with a brand name, or extra service. After determining the value added by each operational activity, it must be compared with the cost of executing that operational activity or the resources consumed by that operational activity. Operational processes and activities that do not create sufficient value to justify their cost should be modified or eliminated.

Approach

When conducting a value enhancement analysis technique, follow these three steps:

- Identifying customer for the program and understanding customer definition of value.
- Categorizing operational activities by the type of value they add.
- Innovating operational processes and activities to maximize value added.

Each of the above steps is elaborated below as it will facilitate an effective application of the value enhancement analysis technique by the business transformation program core team in order to create the needed artifacts. Value enhancement analysis involves the following three steps:

Step 1. Identifying customer for the program and understanding customer definition of value

There may be more than one customer for a program or operational activity, and they can be identified by conducting internal/external customer analysis. This analysis is the process by which those who receive output of such activities (which may be physical goods, services, systems, processes, information, etc.) are identified and analyzed. Understanding the recipients of the output of operational activities determines which activities make an indispensable contribution to providing value.

To identify customers, ask questions such as:

- Who is the ultimate consumer of the company's product or service?
- Who is the recipient of the output from the activity or process cycle?
- What do they expect to receive?
- What output is generated by the activity or process cycle?
- How will they use the output?
- How far beyond the initial customer will the effects of shoddy output be felt?
- How do the consumer needs and wants affect activities along the process cycle?

Step 2. Categorizing operational activities by the type of value they add

Each operational activity in the work execution cycle should contribute as much real value as possible. Value enhancement analysis focuses on producing high-quality goods, services, and processes as quickly and cheaply as possible. It avoids or eliminates rework-type operational activities, as those add to the process cycle time and increase the cost of the output being produced. The activities are categorized based on the kind of value they add.

- *Real value added activity*: An operational activity that adds direct traceable value to the output (product or service) that the end customer is expecting. Most of the time, such an activity is visible to the customer, and there is value in the eyes of the customer. Examples of activities that add real value include those that increase the desirability of a product or service or process resulting from the business transformation program.

- *Business value added activity*: An operational activity required to run the business but that adds no value from the customer's vantage point. Typically, such an activity in a back end activity and it is not visible to the customer. There is no perceived value of that activity in the eyes of the customer, and the activity being seamless could be a reason. Examples include hiring, training, procuring, financial close, and facilities management.
- *Non value added activity*: An operational activity that does not contribute in any direct or indirect way in meeting customer requirements or to running the business and could be eliminated without degrading product or service or process. The transformed future state may not entail executing certain current state operational activities that were put in place ages ago for legitimate reasons at that historical point in time. Certain business changes enabled by the transformation program may also result in redundant operational activities.

Step 3. Innovating operational processes and activities to maximize value added

The categorized operational activities can be further analyzed in group facilitated sessions to generate ideas for significant improvements to them. New processes that lead to breakthrough improvements in real value added can come out of such brainstorming sessions. Answers to the following questions would assist in maximizing the value creation by operational functions:

- Is each real value added operational activity performed as effectively as possible? Can these operational activities be accomplished with lower cost and in shorter execution process cycle times without sacrificing the quality or customer perceptions of value?
- Does each business value added operational activity provide indispensable support for the real value added activities? How can such activities be migrated to the real value added category? What modifications can further improve efficiency, reduce process cycle time, and the cost of conducting these operational activities?

- What options exist to convert the non value added operational activities to either the business value or real value add segments? How can such operational activities be eliminated entirely without damaging the perceived quality or increasing cost?

TRANSITION MODEL OPERATIONALIZATION

The transition of the business transformation program from the program management team to the operations team is a steady process. For a designated agreed period of time, the program team and operations team work in parallel and hand in hand. This approach results in accomplishing the goal of an effective, smooth transition. As exhibited in Figure 11.4, in the first frame, the leadership teams from the program and operations sides are collaborating to ensure alignment, support, and commitment to the goal. The middle frame is a continuum, with the program team owning the beginning and gradually transferring to the operations team as knowledge transfer and on-the-job training ends.

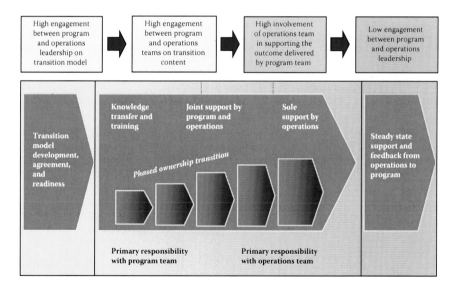

FIGURE 11.4
Transition of program to operations.

WALK THROUGH TECHNIQUE

Overview

This is one of the techniques in the program management tool kit that can be implemented across a wide range of stakeholder types (e.g., internal, external, senior executives, sponsors, or end users) at different junctures of the program to check if an output or process is in line with the expectations. The efficient transition to the operations function can be done through a series of walk throughs. A walk through is a structured meeting held to validate the completion and quality of the output, deliverable, or process. During a walk through, the output is explained ("walked through") in detail by the program management team or owner or subject matter experts to pertinent stakeholders.

Though the walk through technique is being showcased under the "transition to operations" and "sustain outcome delivery" processes of the program management life cycle, it is applied throughout the life cycle. Deliverables like program vision, program strategy, and the program road map have to be walked through with transformation program sponsorship and leadership teams to confirm common understanding and secure buy-in. Examples of process related walk throughs would be transition management, strategic planning, process improvement, change management, systems development, and stakeholder engagement.

Objective

Walk through meetings are conducted to validate any form of deliverable or process. In the case of transition to operations, the knowledge transfer is imperative. One of the avenues for effective knowledge transfer is through walk throughs. Usually, the review with the program sponsors, steering committee, and functional executives is to confirm that the program deliverable or process is matching the requirements and supporting the program business objectives. Upon completion of the walk through, communications to the broader set of stakeholders of the business transformation program is done to increase awareness. As the end goal of a walk through is to secure agreement, commitment, and formal sign-off, it plays an important role in the transfer of responsibility from the program team to the operations team for executing the attained future state.

Approach

Though walk through meetings can be formal, informal, or some combination thereof, the ones with the sponsors and executive team are formal. The steps taken to conduct an effective walk through meeting with the transformation program stakeholders (senior level executives or others) are listed below:

- Determine the walk through meeting objectives
- Socialize meeting objectives and finalize walk through meeting attendees
- Organize and finalize walk through meeting logistics
- Develop and disseminate the "meeting packet" for the walk through meeting
- Lead the walk through meeting
- Complete the post walk through meeting steps

Step 1: Determine the Walk Through Meeting Objectives

The walk through meeting is a formal facilitated meeting with the rigor and structure needed to realize the agreed meeting objectives. One of the first steps to be taken to ensure the success of a walk through meeting is the determination and agreement of the business objectives for the meeting. The program team leading the business transformation program can identify the objectives for the meeting with the help of the inputs of subject matter experts and other stakeholders of the program.

Step 2: Socialize Meeting Objectives and Finalize Walk Through Meeting Attendees

The upfront effort in planning for the walk through meeting will position the meeting to accomplish the intended objectives. The roles and responsibilities of the individuals involved in planning and conducting this meeting should be detailed e.g., logistics planner, facilitator, coordinator, notes taker. The program management team should socialize the walk through meeting objectives with the key stakeholders to get their buy-in. The meeting objectives can be shared at the pertinent governance forums that have been devised for the business transformation program. The program manager should finalize the attendees for the walk through meeting based on the meeting objectives and inputs of the designated subject matter experts.

Step 3: Organize and Finalize Walk Through Meeting Logistics

The walk through meeting logistics and coordination team assists in lining up and managing the logistics support for the meeting. The meeting objectives, attendees, duration, location, and content will influence the logistical needs. It is important to have the right setting for the walk through meeting and the made logistical arrangements should be checked prior to the meeting to ensure readiness of the meeting facility. Also, a confirmation that key needs of all the stakeholder attendees has been factored including their receipt of the "meeting packet" for the walk through meeting is necessary. The precision in logistical arrangements enable the walk through meeting to be productive.

Step 4: Develop and Disseminate the "Meeting Packet" for Walk Through Meeting

The co-developed and finalized meeting objectives will drive the agenda for the walk through meeting. The walk through meeting planning effort aids in the development and creation of the detailed agenda and the supporting "meeting packet" to be discussed at the meeting. The meeting packet contains the materials to be reviewed and analyzed during the walk through meeting should be disseminated to the stakeholders prior to the meeting. As 100% participation of the chosen attendees for the meeting duration is a critical success factor, the meeting invitation should be released with adequate lead time to get that participation. The final meeting packet can be sent to the business transformation program stakeholder attendees a bit closer to the meeting date.

Step 5: Lead the Walk Through Meeting

The program manager (or designee) for the business transformation program leads the walk through meeting. The materials dispatched to the attendees in advance of the meeting is reviewed and discussed at the meeting. Any new materials that may have been developed are also shared at the meeting. Since the right stakeholders are participating and have come prepared for the walk through meeting, the facilitator is positioned to effectively drive the meeting with the help of the agenda and materials. The inputs, questions, ideas, and concerns of the stakeholders that get raised during the walk through meeting need to be addressed to the extent possible during the meeting. The meeting "parking lot" tool should capture those items raised in the walk through meeting that need to be worked off-line or in other meetings. The meeting notes taker documents the meeting notes including the decisions and action items.

Step 6: Complete the Post Walk Through Meeting Steps

The post walk through meeting step entails follow through on the agreed next steps during the meeting. The meeting notes should be compiled, reviewed for completeness and accuracy, and disseminated to the stakeholder attendees. In order to retain and build upon the momentum resulting from the completion of a successful walk through meeting, timely and accurate completion of the action items resulting from the meeting is critical. The program management team needs to be diligent in addressing the items that were put on the meeting parking lot during the walk through meeting. The launch and completion of a walk through meeting feedback survey by the attended stakeholders can provide inputs to the program team on how to further improve future walk through meetings.

LEADING LESSONS LEARNED FROM WALK THROUGHS

The pressure of market forces on organizations to constantly get better and do more with less requires the company leadership to create a culture of continuous improvement. Conducting formal "lessons learned" exercises during the "transition to operations and close program" program management process with multiple stakeholder groups and aggregating the findings from all such exercises provides tremendous intelligence to the organization. The program team analyzes these findings and develops recommendations in working with the stakeholders and subject matter experts. The implementation of the continuous improvement recommendations on a future business transformation program positions the future program and organization for greater business success.

The instrument portrayed in Figure 11.5 can be utilized to lead lessons learned exercises and realize the benefits of such exercises. The best practice is to have the project teams perform such exercises at the project level. Based on the nature and structure of the program, similar exercises

Instrument for Capturing Lessons Learned									
Program Name:									
Project/Track Name:									
Id	Lesson Learned Description	Root Cause Description	Impact	Ranking	Category	Recommendation	Desired Outcome	Approval to Implement	Owner
1									
2									
3									
4									
5									
6									

FIGURE 11.5
Learning from transformation program.

can be performed at the program level to gather insight from all program areas. The outputs of these lessons learned drills can be consolidated and rationalized.

The "ranking" column is the performance scale, e.g., a five-point scale of: exceptional, went well, could have gone better, went poorly, or failed. The category column aids in classifying the lessons learned into groups, e.g., program charter, program plan, external resources, program scope management, program budget, or program status reporting. An analysis of these completed drills will result in common improvement themes across projects, specific improvements for certain projects, and improvements at the program level. The benefit and desired outcome from implementing these improvements or recommendations should be called out to secure approval for implementing. An owner to oversee implementation of improvements in a future program is designated.

BUSINESS OUTCOME DELIVERY AND SUSTAINMENT

The success of a program designed and launched to transform the entire business or a major part of the business hinges on the attainment of the desired future state, achievement of the strategic business objectives, and delivery of the desired business outcomes. Transformation program leadership, sponsorship, stakeholder expectation management, rigorous monitoring, and smooth transition to operations dictate the delivery and sustainment of business outcomes.

Business Outcome Delivery Analysis								
Id	Business Objective	Outcome Description	Key Value Indicators	Outcome Realization Owner	Outcome Realization Target Date	Outcome Realization Status	Realized vs. Expected Outcome	Plan to Close Gaps in Outcome	Action Plan for Sustaining Outcome
1								
2								
3								
4								
5								
6								

FIGURE 11.6
Program outcome delivery analysis.

Figure 11.6 exhibits a valuable tool to not only assess the initial program success on completion of the transformation program, but to also monitor the ongoing success through outcome sustainment. Strategic business objectives and corresponding business outcomes are captured. The key value indicators (KVIs) and the underlying value metrics associated with each business benefit are noted. The owner accountable for targeted outcome delivery and the time frame for outcome delivery are listed. The actual and targeted outcome realization can be compared to identify gaps in outcome delivery. Performance of root cause analysis will identify the actions needed to close the gaps, and the owner accountable for outcome delivery will drive completion of these actions.

The successful completion of a transformation program results in business outcome delivery in the near term, but that is only a partial success. Business outcome sustainment over time by the operations team leads to full success. The acceptance of business change and embrace of the change by the operations and other stakeholders will influence the delivery of business outcomes. Often, large programs that are innovating or renewing the business require cultural shifts to operate in the attained future state. Ultimately, such culture shifts will drive sustainment of the business outcomes. Performance evaluation and incentive reward systems for operational functions should be reviewed to ensure their alignment to business outcome delivery and sustainment. The implementation of continuous improvements based on the insights gathered from value enhancement analysis, administration of periodic customer satisfaction surveys, and completion of lessons learned exercises will also facilitate business outcome delivery on an ongoing basis.

Case Study: Strategic Initiative to Get and Stay Compliant with Government Regulations

CONTEXT

A well established U.S.-based Fortune 500 company, which is the largest health benefits solution provider, was challenged in complying with the health plan pricing and health insurance renewal regulations put forth by the CMS (Centers for Medicare and Medicaid Services) regulatory body. This public company with nationwide operations had a comprehensive health benefits solution portfolio comprising certain niche solutions that targeted certain underserved customer segments. The highly successful company enjoyed superior profit margins on such niche solutions.

BUSINESS PROBLEM OR OPPORTUNITY

The CMS formally informed the health benefits solutions company of their intent to revoke the company's license to sell some of these niche market health benefit solutions within twelve months if the company was not able to provide adequate evidence of complying with the customer disclosure requirements surrounding health insurance premium changes and insurance policy renewals. Some of the company's customers had complained to the CMS about the company's practices. The company had been experiencing pain points in the issuance of accurate, timely disclosures to its customers who were spread across the nation in different states. As the nature of disclosure and the notice period requirements varied by state and health benefits solution, and since multiple functions were involved in the issuance process, the complexity was high. The cross functional coordination amongst many internal stakeholders (products, actuary, systems, compliance, legal, enrollment, servicing, and renewal) and external stakeholders (customer, CMS, print vendors) was needed.

SOLUTION

A formal strategic initiative was launched to evaluate and transform the company's operational practices and processes. A centralized shared services group was created to plan, execute, and monitor the disclosure requirements for all products, U.S. states, and customers. The program team assigned to the strategic initiative developed a transition model to facilitate an effective transition of ownership of the disclosure function to this central group. The engagement process with CMS was streamlined to facilitate proactive and effective communications. The program management team partnered with functional subject matter experts in redesigning germane business processes and removing non value added activities by performing value enhancement analysis. As part of the phase of transition model execution, the shared services group participated in a series of training sessions and knowledge transfer walk throughs in getting ready to execute.

BUSINESS OUTCOMES AND BENEFITS

The health benefits solution provider furnished timely and adequate evidence to the CMS of the actions taken and the changed business practices that adhered to all the regulations. The business process and systems controls to proactively detect and rectify any noncompliance activity were implemented. The enhanced processes and systems resulted in higher throughput from the print vendors and improved internal productivity. The creation and implementation of an operations transition plan is critical to delivering business outcomes and sustaining benefits realization.

SUMMARY

The timely roll-out of the end-to-end operations transition framework will position the designated operations functions to move forward the charter of the business transformation program once it ceases to exist. The readiness of the operations team is facilitated by the development and execution of the program transition plan, which comprises the operations plan, training plan, and knowledge transfer plan. The value enhancement analysis technique promotes operational excellence in everything the operations group does to realize and sustain the desired business outcomes. The pilot approach in transitioning from the program mode to an ongoing business operations mode mitigates risks. The gradual transition of knowledge, ownership, and responsibilities can be done through the technique of walk throughs. The embedment of continuous improvement practices and lessons learned exercises creates an operations environment that will sustain the delivery of expected business outcomes.

REFERENCE

PMI. 2014. *Pulse of the profession in-depth report: Enabling organizational change through strategic initiatives*. Newtown Square, PA: Project Management Institute. http://www.pmi.org/~/media/PDF/Publications/Enabling-Change-Through-Strategic-Initiatives.ashx

12

Executive Summary

A business transformation program aligns strategies and critical business processes to create a competitive advantage through higher revenue, higher earnings, and reduced operational costs. For organizations seeking a competitive edge, program management is the "secret sauce" in achieving transformation program objectives, realizing program outcomes, delivering business results, and setting the stage for ongoing benefits realization. As organizations become more global and their initiatives more complex, effective program management will be critical to achieving strategic objectives and realizing successful business outcomes.

Program management will continue to play a key role in closing the business outcome gap. Program management is the glue that brings strategy, people, process, technology, structure, and measurement dimensions together to attain the future state expected of a transformation initiative. Figure 12.1 depicts a holistic framework that leverages the strength of program management in effectively integrating multiple disciplines and methods. This program management driven framework can be capitalized by enterprises to successfully execute a complex transformation program that delivers the future state of greater competitive advantage in the marketplace.

The following topics are revisited as an executive summary in this final chapter of the book:

- Road map for transformation (or strategic initiative) success
- Summaries of Chapters 2–11
- Revisit of main points
  - Program architecture: Bridge to implementing business strategy
  - Business outcome and benefits realization life cycle management
  - Program management life cycle
  - Program management office and office of business transformation
- Key takeaways on program management

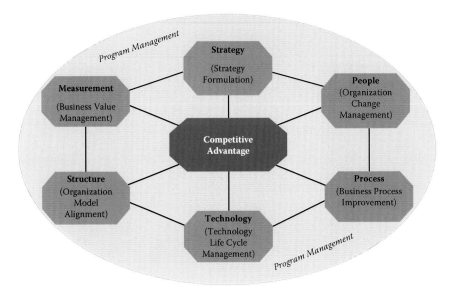

FIGURE 12.1
Program management enables improved competitive positioning.

ROAD MAP FOR TRANSFORMATION (OR STRATEGIC INITIATIVE) SUCCESS

Established program managers with highly successful track records apply their "hard" program management skills (e.g., integration management, scope management, cost management, and time management) as well as their "soft" program management skills (e.g., consensus building, expectation management, decision framing, coaching, and motivating) to drive a business transformation program to success. Program managers oversee the process of solving complex business challenges by driving end-to-end processes, managing project managers who oversee project work streams, and engaging with stakeholders to manage their expectations of program outcomes.

True transformation program management—and, ultimately, program success—is best achieved by flawless execution of the following road map:

- Success starts upfront: Describe the problem accurately (Chapter 2)
- Articulate the program vision and objectives (Chapter 3)
- Secure cross functional executive sponsorship (Chapter 4)

- Develop and implement a governance model (Chapter 5)
- Define success, outcomes, and key value indicators (Chapter 6)
- Invest in planning and creating an integrated approach (Chapter 7)
- Drive strong partnership and stakeholder engagement (Chapter 8)
- Provide leadership across all levels (Chapter 9)
- Monitor aggressively and have contingencies (Chapter 10)
- Create and implement an operations transition plan (Chapter 11)

Each of the ten components of this road map—the steps needed to make a business transformation effort attain the future state objectives—has been allocated a chapter in this book. Though the road map components are strategically sequenced, the planning and execution of the road map is an iterative exercise. Figure 12.2 presents a pictorial view of the proven road map to drive a business transformation program to success in the eyes of the stakeholders. The program management team needs to have the capability (hard skills, soft skills, methods, processes, techniques, and tools) to execute against the road map. The program management team also utilizes the techniques in the program management tool kit to deliver in line with the road map.

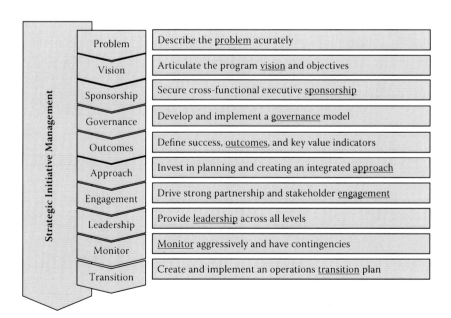

FIGURE 12.2
Road map for transformation (or strategic initiative) success.

Per PMI's *Pulse of Profession In-Depth Report: Navigating Complexity* (2013), which surveyed 697 project management practitioners, high-performing organizations (those that achieve 80% or more of projects on time, on budget, and meeting original goals) have much higher levels of program management maturity. Some of the key program management techniques have been expounded in this book. The following section provides an executive-level summary snapshot of each of the ten road map components (as presented in Chapters 2–11, respectively).

SUMMARIES OF CHAPTERS 2–11

Chapter 2: Success Starts Upfront: Describe the Problem Accurately

The program management team plays a critical role during the program frame-up phase by ensuring that there is due diligence around business problem definition and alignment among the stakeholders on the problem that the business transformation program will solve. The environment scanning and voice-of-customer techniques assist in identification, definition, and validation of the business problem. Effective initial and ongoing communications are needed to reinforce the business problem the transformation program has been designed to solve. Clarity around the problem statement, strategic business objectives, and desired business outcomes is a "must have." A common understanding and agreement of the problem statement among the stakeholders creates an invaluable platform that the program management team can further build upon as they successfully drive the program forward.

Chapter 3: Articulate the Program Vision and Objectives

Programs that are designed to transform a business are complex and high risk, as the magnitude of change being driven is high. As the program management team drives forward the planning and execution of programs, all of the stakeholders need to be cognizant of the program vision, program strategy, and program objectives and goals. In addition to awareness of the program vision, for true business change to happen at all levels, there has to be a buy-in to the vision. The strategic risk of program execution without comprehending program vision is similar to the risk of program execution without adequate program planning. Periodic reinforcement of the program vision and how the achievement of the program objectives will benefit everyone is one of the most important critical success factors for a transformation program.

Chapter 4: Secure Cross Functional Executive Sponsorship

The sponsorship team needs to own the transformation program and be visibly involved. The drivers for the business transformation need to be reinforced by the sponsors. The sponsor alignment to the program strategy, vision, objectives, and outcomes needs to be evident to the program stakeholders. Explicit periodic communication of the sponsors' commitment to the transformation program will support the cause. A cross functional and multilevel sponsorship model is needed to authorize and legitimize the program. The program's value justification technique provides the fact, data, and objective based business case for the program to keep marching ahead until the attainment of the program mission. The proactive, upward management of the sponsor by the program management team is key for the program team to get the continued needed support. By addressing the program barriers, the sponsorship team positions the program team to realize the desired business outcomes expected of the transformation program.

Chapter 5: Develop and Implement a Governance Model

For a transformation program to succeed, it is imperative to develop and implement the governance practices sooner rather than later. A program governance model is a combination of governing bodies, strategic control and oversight functions, and cohesive policies that defines the consistent management of the program throughout the program life cycle. The key constituents of the program governance model are program organization structure, governing body definition, roll-out of governance forums, program accountability, escalation paths, and decision making processes.

Program governance ensures that strategic direction and program vision are aligned, that program priorities are defined and understood, and that decisions are aligned with the overall business objectives. A lack of robust governance practices poses substantial risk to realizing the desired business outcomes targeted by the transformation program. The implementation of the SMART principle facilitates governance around the program delivering the benefits expected by key stockholders on attainment of the future state.

Chapter 6: Define Success, Outcomes, and Key Value Indicators

The program manager has to ensure that the definition of program success, the business outcomes, and the key value indicators (KVIs) to assess program success follow the SMART principle. These need to be revisited at key program milestones throughout the program life cycle to confirm continued alignment. Such a proactive approach is similar to the construct of program risks needing to be identified, mitigated, and watched from program initiation through program closure. The importance of driving adequate clarity upfront on what constitutes the success of a transformation program is high.

Elaboration of the expected outcomes and timing the delivery of those outcomes are critical factors in the success of the program. Equally important is the periodic communication and education around the definition of a successful transformation program, targeted business outcomes, performance improvement measurement architecture, and KVIs to all stakeholder groups. Program management has to constantly manage stakeholder expectations and validate that there are no deviations from the original definition of success, the strategic criteria to judge success, and the KVIs. The activities associated with management of the business outcome life cycle have to be planned in the integrated transformation program plan, and these activities have to be executed and monitored.

Chapter 7: Invest in Planning and Creating an Integrated Approach

Based on the key transformation program planning concepts and covered processes, one can see that program planning requires significant effort and acumen. The five-step program architecture process establishes the bridge between business strategy and the implementation of that strategy, and successful completion of implementation results in benefits realization. Program vision aligns the program to the strategic direction of the enterprise. The output of the program architecture is the key input to the program management life cycle. The program strategy and program road map drive the transformation program plan to ensure continued alignment. The transformation program planning technique assists in the development of a comprehensive program plan, which has numerous component plans, including the business outcome realization plan.

The bottom up approach of creation of the project plans and higher level aggregation of those in the program plan facilitates the buy-in of the core program team, which does the heavy lifting throughout the program management life cycle. As the execution of the baselined integrated program plan shifts the organization from the current to the desired future state, the program leadership team needs to proactively ensure organizational and individual readiness and acceptance to the strategic change. In other words, the "people" dimension needs to be accounted for by the transformation program. The incorporation of organization change management practices throughout the program management life cycle facilitates organization readiness and other behavioral components. An end-to-end holistic and integrated approach will result in attainment of the future state, accomplishment of the strategic objectives, and sustainment of the defined business outcomes.

Chapter 8: Drive Strong Partnership and Stakeholder Engagement

The lack of a comprehensive stakeholder engagement strategy and execution of that strategy along with the rest of the work driven by the program management life cycle is a recipe for failure of the business transformation program. Fusion of the organization change management plan into the overall program plan lays the platform for garnering stakeholder expectation management, stakeholder engagement, and stakeholder commitment to the transformation agenda of the program. Communication, change readiness, training, and coaching not only enable stakeholder expectation management, but they also support the change process in transitioning the organization, business unit, or function from the current state to the improved future state.

One of the consistent continuous improvement themes seen in the postmortem exercises of completed programs is the opportunity for the program team to better engage with the stakeholder community. The program executive sponsor has to regularly engage with executive level stakeholders and champion the program. For a transformation program to successfully deliver and sustain the strategic change, the core program team has to proactively and effectively partner and engage with the identified stakeholders.

Chapter 9: Provide Leadership across All Levels

The complexity of business transformation programs requires cross functional, cross dimensional, and cross project leadership to realize the program vision and sustain business outcomes. The significant business change initiated by a transformation program heavily impacts a large number of stakeholders at different levels and affects the organization as a whole. Leadership in analyzing and communicating the impact, reinforcing commitment to the needed support, and sharing the benefits of transformation program is paramount. Leadership over the development, implementation, and communication of a comprehensive support strategy to counter the change impact is essential to increase the acceptance rate of the program among the impacted parties.

Leadership behavior has to permeate through multiple levels of the transformation program organization and not just at the level of the executive sponsor or the steering committee. Leadership on the technical and human side is essential for the entire duration of the program management life cycle. The office of business transformation provides an integrated leadership model that facilitates getting the right quantum of leadership to all of the dimensions of program management. Program leadership—or a lack of it—can make or break a program devised to transform a business.

Chapter 10: Monitor Aggressively and Have Contingencies

Institutionalization of program controls is essential for efficient operational execution of the business transformation program and to position the program for success. The program management life cycle process of "monitor program delivery" interjects these controls and assists with effective management of the strategic and tactical risks to the business transformation program. The timely and accurate capture, analysis, and communication of the program status to the program stakeholders aids in timely interventions for the program to stay aligned to the plan. Actions needed to correct the course are determined by the monitoring processes, and implementation of the approved actions happens through the program execution processes.

Dashboards showing the status of transformation progress at the program and project levels are used to keep stakeholders informed of the program's status and manage stakeholder expectations. Monitoring of program milestones validates that the program timeline is in line with the integrated program plan. Program scope validation confirms that the produced outputs are in line with the charter, and program financial tracking ensures that budget parameters are met. The issues and risk management tools support the implementation of the critical issues and risk management processes. The program monitoring process drives additional value into the business transformation program by creating opportunities for continuous improvements. Finally, tested contingency plans mitigate the significant risks encountered by the transformation program.

Chapter 11: Create and Implement an Operations Transition Plan

The timely roll-out of an end-to-end operations transition framework will position the designated operations functions to move forward the charter of the business transformation program once it ceases to exist. The readiness of the operations team is facilitated by the development and execution of the operations plan, training plan, and knowledge transfer plan. The value enhancement analysis technique promotes operational excellence in everything the operations group does to realize and sustain the desired business outcomes. The pilot approach in transitioning from program mode to ongoing business operations mode mitigates risks. The gradual transition of knowledge, ownership, and responsibilities can be done through the technique of walk throughs. The embedment of continuous improvement practices and lessons learned exercises creates an operations environment that will sustain the delivery of expected business outcomes.

REVISITATION OF MAIN POINTS

The following key themes across the various chapters are emphasized in this section:

- Program architecture: Bridge to implementing business strategy
- Business outcome and benefits realization life cycle management
- Program management life cycle
- Program management office and office of business transformation

Program Architecture: Bridge to Implementing Business Strategy

The enterprise vision, mission, strategy, values, and strategic objectives can be used as the foundational elements to commence designing the transformation program. Figure 12.3 is an illustrative elaboration of the approach the program manager takes to design the program architecture. After the program vision has been formulated, the business problem is analyzed as part of a current state assessment; a blueprint of the desired future state is next developed; and finally the transformation program is designed. The seamless integration between program architecture and program management life cycle results in the end-to-end approach to implement the strategy to transform the business at an enterprise or business unit or functional level based on the scope.

Business Outcome and Benefits Realization Life Cycle Management

Program management is the "secret sauce" to deliver business outcomes and realize benefits as it orchestrates the process of setting clear, measurable,

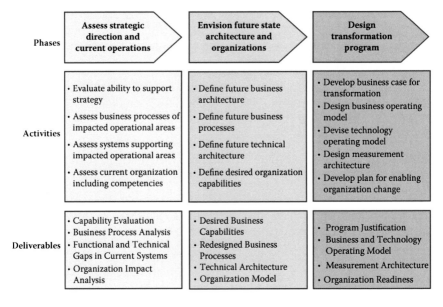

FIGURE 12.3
Approach to designing program architecture.

and achievable business outcomes and benefits for the program. Business outcome modeling defines and prioritizes the business outcomes that a transformation program is chartered to deliver. Enterprises with higher success rates on their transformation programs do not look at strategic objectives, business outcomes, and benefits just at the start and end of the program. The best practice is to approach program business outcomes and program benefits as a continuum and adopt a life cycle approach to managing them. The life cycles for business outcome delivery and benefits realization have to be embedded in the management life cycle of the transformation program. Program management drives the end-to-end process beginning with business outcome and benefits definition and ending with realization and sustainment of business outcome and benefits.

Program Management Life Cycle

The completion of program architecture work triggers the commencement of the program management life cycle. The program management life cycle plans and implements the business transformation program. It also effectively transitions the future state attained by the program to the operations function, which will own the ongoing execution and business outcome realization. The eight processes comprising the program management life cycle brings transformation to fruition by driving the steps needed to transition the business from the current state to the future state. The "develop a program road map" process crafts the phased implementation plan after working through strategic implementation scenarios and analyzing their feasibility. The program road map to get to the desired future transformed state typically encompasses sub-road maps, e.g., people road map, process road map, and technology road map.

The "create the program plan" process builds out the comprehensive transformation program plan detailing the work to be performed under the program. The program plan brings together the six program management dimensions, program management life cycle phases, and program management processes. The integrated program plan enumerates a holistic approach to solve the defined business problem the business transformation program owns by accounting for all the six dimensions—strategy, people, process, technology, structure, and measurement. Figure 12.4 exhibits the end-to-end program management model consisting of the program architecture and the program management life cycle.

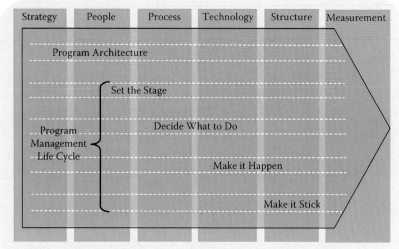

Multi-dimensional and Integrated Program Management Model

FIGURE 12.4
Program management model: secret sauce to deliver outcomes.

Program Management Office and Office of Business Transformation

The framework of "think–design–build–operate" can be employed to develop and implement the structure needed to effectively plan and execute the business transformation program. The program management office (PMO) for the transformation is the overarching structure to provide strategic execution leadership on the program. Under the leadership of the program management team, the office of business transformation (OBT) is structured under the transformation PMO. The OBT allows for integration of the people, process, and technology capabilities into the program management capabilities and practices. OBT is part of the PMO, and it comprises the organization's change management office, the office of business process innovation, and the technology management office. The PMO is a conduit for employing program management, people, process, and technology best practices through the journey of the business transformation program.

KEY TAKEAWAYS ON PROGRAM MANAGEMENT

- Program management is the vehicle that enterprises use to implement business strategy and accomplish their strategic objectives, which results in improved competitive positioning.
- Program management is the glue that brings strategy, people or human capital, process or business operations, technology, organizational structure, and measurement dimensions together to drive a successful business transformation.
- Program management develops the program strategy, program road map, and integrated program plan and then executes and monitors the integrated program plan rigorously to enable the success of a business transformation program.
- The program management operating model, which maps the solution life cycle to the six dimensions, is a thorough framework for defining and solving complex business problems.
- Program leadership and program management are complementary, and both are needed to successfully execute strategic programs that deliver tangible top and bottom line business results and additional intangible benefits.
- A program management tool kit employs techniques that enterprises use to successfully deliver the outcomes expected of a complex, strategic, cross functional business transformation program.
- Program management well integrates the business and technology sides in transitioning an organization from its current to its envisioned future state as part of the transformation initiative.
- Although program management is not a silver bullet, it can play a significant role in closing the gap in stakeholder minds between the expected and realized business outcomes.

CONCLUSION

Enterprises need to constantly innovate business models, products, services, processes, and systems to maintain a strong competitive position in the marketplace. Enterprises design and launch business transformation programs to drive such innovation. The typical financial outcomes expected in the future state attained by the transformation program are increased revenues, improved earnings, and reduced costs. For business transformation programs to succeed in today's ever changing market environment, enterprises must give renewed emphasis to the tenets of program management.

A holistic, end-to-end program management practice is critical to make a business transformation happen and sustain. The program management discipline integrates and aligns the six critical dimensions (strategy, people, process, technology, structure, and measurement) needed to transform a business through successful completion of a transformation program. Program management recognizes that organizational change is required to realize and sustain the business outcomes targeted by the transformation program. The execution of the integrated program plan drives and readies the organization to handle the business change.

Program management is a key enabler for improving the success rate of business transformation programs. Enterprises that invest in bolstering their program management capability and maturity levels are at a strategic advantage. Program management improves the organization's performance level and mitigates the enterprise's risk of poor implementation of formulated business strategies. Competencies in program management are a powerful competitive weapon, as they enhance the strategic execution capabilities of enterprises.

REFERENCE

PMI. 2013. *Pulse of the profession in-depth report: Navigating complexity*. Newtown Square, PA: Project Management Institute. http://www.pmi.org/~/media/PDF/Business-Solutions/Navigating_Complexity.ashx

Vocabulary

B

baseline (n. and v.)
benefits realization (adj.)
build out (n.)
business change (adj.)
business outcome (adj.)
business problem (adj.)
business transformation (adj.)

C

change control (adj.)
change management (adj.)
closeout (n.)
cloud computing (adj.)
continuous improvement (adj.)

F

frame up (n. and adj.)

H

health care (adj.)

K

knowledge transfer (adj.)

L

life cycle (adj.)

M

multidimensional
multilevel

O

operationalization
operationalize
operations transition (adj.)
organization change (adj.)

P

performance improvement (adj.)
posttraining (adj.)
program governance (adj.)
program management (adj.)
program monitoring (adj.)
program planning (adj.)
program transition (adj.)

R

road map (n.)
roll-out (n.)
roll-up (n.)

S

sign-off (n.)
sign off (v.)
strategic initiative (adj.)
switchover (n.)

T

takeaway (n.)
time frame (n.)
timeline (n.)
tool kit (n.)
transformation program (adj.)

U

UK/US (when abbreviation is appropriate)

V

value enhancement (adj.)
value justification (adj.)

W

walk through (n.)
workforce (n.)

Index